On the Move

Translating Cultures in the Early Modern World

VOLUME 2

The titles published in this series are listed at *brill.com/tcemw*

On the Move

Mobility and Early Modern Translation

Edited by

Giancarlo Casale
Ann Thomson

BRILL

LEIDEN | BOSTON

Cover illustration: A Dragoman of the Ottoman Divan, circa 1700. From Cristoforo Conte Mamuca della Torre (himself a dragoman), *Figurae 294 colorite che rappresentano le cariche, le professioni ecc. della città di Constantinopoli e di varie nazioni barbare colla descrizione.* 3 vols. Österreichische Nationalbibliothek (ÖNB), Cod. 8562–8564, vol. 1, fol. 198[a]. Public domain image, courtesy of the ÖNB.

The Library of Congress Cataloging-in-Publication Data is available online at https://catalog.loc.gov

Typeface for the Latin, Greek, and Cyrillic scripts: "Brill". See and download: brill.com/brill-typeface.

ISSN 2772-8331
ISBN 978-90-04-73179-0 (hardback)
ISBN 978-90-04-73180-6 (e-book)
DOI 10.1163/9789004731806

This book is printed on acid-free paper and produced in a sustainable manner.

Contents

Preface

Several of the contributions to this volume originated in a workshop on 'Translators, Translations and Mobility in the Mediterranean', held at the European University Institute on 31 January and 1 February 2022. This workshop was part of the COST Action (CA. 18140) 'People in Motion: Entangled Histories of Displacement across the Mediterranean (1492–1923)' (PIMo), funded by the European Union and chaired by Giovanni Tarantino (University of Florence). This publication, centered on the question of translation and mobility, extends the geographical area concerned beyond the Mediterranean, while still keeping Mediterranean preoccupations at the heart of the investigations. We would like to thank those participants in the PIMo workshop who were unable to contribute to the present publication (Elisabetta Benigni, Ida Caiazza, Jorge Flores and Elisavet Papalexopoulou) for their contributions to the discussion on that occasion.

Figures

Notes on Contributors

Giancarlo Casale
is an author and translator specialising in the history of the Ottoman empire and its relations with the larger early modern world. He is currently professor of early modern Mediterranean history at the European University Institute in Florence.

Alessia Castagnino,
who has a PhD in European History from the University Ca' Foscari of Venice (2014), is currently a Lecturer in Early Modern History at the University of Florence and Principal Investigator of the project 'NEGOTIATINGNEWS. Negotiating and disseminating medical knowledge in the eighteenth-century Italian periodical press'. Her main research interests concern the intellectual and cultural history of the Enlightenment, the social and cultural history of translation, and the application of digital humanities in historical research. Among her recent publications: 'Anglo-Italian cultural relations "through the lens of translation." The first Italian editions of William Robertson's *History of Scotland*' in *Ideas across Borders: Translating Visions of Authority and Civil Society in Europe* ed. G. Mahlberg, T. Munck (2024); 'Status and agency of translators in early 19th century Italy: Recent studies and critical insights', *Contemporanea* 37, 2024.

Angelo Cattaneo
is Principal Investigator for the CNR, Rome and Adjunct Professor at the University of Florence. His research focuses on the history of mapping, travel literature and intercultural encounters, in particular on missionary practices, linguistics and mapping, at the interface of European and Asian empires (1250–1700). He is the P.I. of the PRIN 2022 Research Project 'Mapping and Translating Spaces, Cultures and Languages. Experiences from the Missions connected to the Portuguese Empire (1540–1700)'. His publications include *Fra Mauro's* Mappa mundi *and Fifteenth-Century Venice* (2011), *Tradurre il mondo. Le missioni, il portoghese e nuovi spazi di lingue connesse* (2022) and several edited volumes. In 2024 he was appointed Visiting Professor at Yale University.

Tunahan Durmaz
is a Ph.D. candidate in the Department of History at the European University Institute, Florence. He studies the social, cultural and political aspects of medical knowledge in the early modern Ottoman world. His dissertation, tentatively titled 'Mood, Appetite, and Sores: Medicalization of Illness in Otto-

man Istanbul, 1640–1691', explores perceptions of disease and illness in late seventeenth-century Ottoman Istanbul by tracing the medicalisation of mood, appetites and sores in contemporary medical texts.

Stephan Hanß
is professor of early modern history at the University of Manchester. His recent books include *Narrating the Dragoman's Self in the Veneto-Ottoman Balkans, c. 1550–1650* (Routledge, 2024).

Giovanni Lista
is a Fritz Thyssen Fellow at IZEA, Martin-Luther-Universität Halle-Wittenberg, Germany. He obtained his PhD in intellectual history at the European University Institute and has held postdoc positions at the Lichtenberg-Kolleg (Göttingen) and at the Herzog August Bibliothek (Wolfenbüttel). He has published on seventeenth-century republicanism, natural law and empires. He has contributed extensively to the biographical dictionary of European translators (EUTEC) and has since developed an interest in history of knowledge, cultural history of science and translation studies.

José María Pérez Fernández
teaches early modern cultural history, translation studies, and comparative literature at the University of Granada in Spain. He conducts research in these disciplines both at home and abroad, as a visiting fellow at institutions that include the European University Institute, Villa I Tatti in Florence, as well as the Universities of Oxford and Cambridge. (https://josemariaperezfernandez.academia.edu/)

Baki Tezcan
is professor of history at the University of California-Davis. He is a specialist of pre-modern Middle Eastern history. His works include *The Second Ottoman Empire: Political and Social Transformation in the Early Modern World* (Oxford, 2012).

Ann Thomson
is emerita professor of intellectual history at the European University Institute. She has published widely on the long eighteenth century in Europe, particularly on the 'Natural History of Man(kind)', the circulation of ideas and information, intellectual networks and translation.

Introduction

Ann Thomson

As Susan Bassnett has pointed out, 'travelling and translating are not transparent activities. They are definitely located activities, with points of origin, points of departure and destinations'.[1] This is of course a reminder from a specialist in translation studies to reflect on the embeddedness of translation, and to go beyond the simple comparison of a source text and its translation, by thinking about the cultural and material aspects of the activity of translation. The remark also reflects the link between translating and travelling that has increasingly been highlighted in recent studies, and more specifically the relationship between travel writing and translation.[2] Translation is, in addition, often used as a metaphor, sometimes for aspects of individual mobility and writing about it and at others for the travel writer's depiction of the country described.[3]

The present volume, however, takes a different approach to the link between travel and translation. This collection is not concerned with travel writing; instead, it brings together historical studies of different aspects of physical mobility as it relates to both translators and texts, in a way which has, perhaps surprisingly, been relatively neglected so far. The lack of interest in the mobility of those who translated reflects the relative lack of interest for long shown concerning the individual translator, beyond specific cases of well-known writers. Translation studies tended until recently to centre on the texts themselves and the complex operation of translation rather than on the individuals who carried out this operation. Historians, on the other hand – even intellectual historians interested in questions of dialogue and intellectual traditions – have only relatively recently become interested in the question of translation and its vital role in transfers and the construction and circulation of knowledge. For long these questions were marginal to the interests of most historians, although they did appear in studies looking, for example, at the activity of

1 'Constructing Cultures: The Politics of Traveller's Tales', in *Comparative Literature: A Critical Introduction* (Oxford: Blackwell, 1993), 114.

2 For example: Michael Cronin, *Across the Lines: Travel, Language, Translation* (Cork: Cork University Press, 2000); Loredana Polezzi, 'Translation, Travel, Migration', *The Translator* 12:2 (2006): 169–188; Susan Bassnett, 'Translation and Travel Writing', in *The Cambridge History of Travel Writing*, ed. Nandini Das and Tim Youngs (Cambridge: Cambridge University Press, 2019), 550–564.

3 See for example Mirella Agorni, *Translating Italy for the Eighteenth Century. Women, Translation and Travel Writing 1739–1797* (Abingdon: Routledge, 2002).

 | DOI:10.1163/9789004731806_002

Christian missions in China.[4] The increase in interest in translation among historians – to which this volume aims to contribute – has of course been stimulated by the rise of studies of cultural transfers, first developed in connection with the elaboration of a German identity during the *Sattelzeit*[5] and then extended to a much broader field.[6] The study of cultural transfers was originally concerned to highlight the important role played by translation in the development of national cultures and identities; as Stefan Berger and Peter Lambert put it, the study of cultural transfers 'seeks to demonstrate the extent to which such national cultures depend on a dialectical process through which indigenous and foreign elements are selectively appropriated.'[7] With the more recent emphasis on connected history, the field has since developed to look more broadly at the interaction between different cultural areas.[8] Even here, however, the person of the often humble translator has not always attracted the attention it deserves. The invisibility of translators, famously discussed by Lawrence Venuti,[9] has generally applied not only to their translations, but also to their existence as historical actors. However, as Stefanie Stockhorst puts it, 'the transfer dimension of translation processes has evolved into a key issue both with respect to intercultural communication and to the constitution and shift of meaning in cultural exchange'.[10] The ensuing recent emphasis on the study of 'cultural translation'[11] has encouraged interest in the role of the trans-

---

4 See R. Po-Chia Hsia, 'The Catholic mission and translations in China, 1583–1700', in *Cultural Translation in Early Modern Europe*, ed. Peter Burke and R. Po-Chia Hsia (Cambridge: Cambridge University Press, 2007), 39–51.

5 *Transferts. Les relations interculturelles dans l'espace franco-allemand* (*XVIIIe et XIXe siècle*), ed. Michel Espagne and Michael Werner (Paris: Editions Recherche sur les Civilisations, 1988); Michel Espagne, *Les transferts culturels franco-allemands* (Paris: P.U.F., 1999).

6 For example: *Cultural Transfer through Translation. The Circulation of Enlightened Thought in Europe by Means of Translation*, ed. Stefanie Stockhorst (Leiden: Brill, 2010); *Cultural Transfers: France and Britain in the long eighteenth century*, ed. Ann Thomson, Simon Burrows and Edmond Dziembowski (Oxford: Voltaire Foundation, 2010).

7 Stefan Berger and Peter Lambert, 'Intellectual Transfers and Mental blockades: Anglo-German Dialogues in historiography', in *Gedächtnis im deutsch-Britischen kulturellen Austausch 1750–2000*, ed. Stefan Berger, Peter Lambert and Peter Schumann, (Göttingen: Vandenhoeck & Ruprecht, 2003), 12.

8 Michael Werner and Bénédicte Zimmermann, 'Beyond comparison: *histoire croisée* and the challenge of reflexivity', *History and Theory* 45 (2006): 30–50.

9 Lawrence Venuti, *The Translator's Invisibility: A History of Translation* (London: Routledge, 1995).

10 *Cultural Transfer through Translation*, 23.

11 For a recent discussion, see Gaby Mahlberg and Thomas Munck, 'Introduction' in *Ideas Across Borders: Translating Visions of Civil Society and Authority in Europe ca. 1600–1840*, ed. Gaby Mahlberg and Thomas Munck (London: Routledge, 2024), 1–18.

lator: the first in Peter Burke's list of the 'six large questions' to which his study of the 'cultures of translation' attempts to provide answers is: 'who translates?'[12] The answer to this question is however not always easy, apart from a number of high profile writers and translators, and while recent research has brought out of the shadows a certain number of these actors, much remains to be done, and a certain amount of caution is required.[13] As Mirella Agorni points out,

> we must be aware of the danger of approaches that move rewriters to the centre of the stage, without simultaneously providing a detailed socio-cultural contextualization of their historical role. Attention to individual agency has to be balanced against attention to the cultural dynamics that have shaped it – and have at the same time been shaped by it. Rewriters are not entirely autonomous, self-directing individuals, but negotiate and produce discourse, while, at the same time, their agency is itself a product of such negotiations.[14]

Although a start in such socio-cultural contextualisation has been made, particularly for Europeans, the social history of translators is still in its early stages.[15] The professionalisation of translation really only developed widely in the nineteenth century, and those who translated over the centuries came from a variety of professions (members of religious orders of course, like Antonini, Vestrini and Antoniutti, who appear here in the chapters by Lista and

12 *Cultural Translation in Early Modern Europe*, 11.

13 See for example: Jean Delisle and Judith Woodsworth, *Translators Through History* (Revised Edition, Amsterdam/Philadelphia: John Benjamins, 2012), and the ongoing online biographical dictionary of European translators in the long Eighteenth Century: https://www.eutec-project.it/. There are also numerous studies of individual writers, as well as attention paid to communities of translators, such as: Patrick Leech, *Cosmopolitanism, dissent, and translation. Translating radicals in eighteenth-century Britain and France* (Bologna: Bonomia University Press, 2020); Alessia Castagnino, 'Il ruolo di mediazione culturale dei traduttori. Un contributo allo studio delle attività dei Gesuiti nel Settecento' (PhD diss., Università di San Marino, 2020).

14 Agorni, *Translating Italy*, 4–5.

15 See Fritz Nies, 'Rhétoriciens de métier, pleins d'aplomb. Vers une histoire sociale des traducteurs français', *Francia. Forschungen zur westeuropäischen Geschichte* 38 (2011): 229–235; *Histoire des traductions en langue française XVIIe et XVIIIe siècles*, ed. Yves Chevrel, Annie Cointre, and Yen-Mai Tran-Gervat (Paris: Verdier, 2014), ch. 2; *Transkulturalität nationaler Räume in Europa (18. Bis 19. Jahrhundert) / La transculturalité des espaces nationaux*, ed. C. Charles, H.J. Lüsebrink, Y.G. Mix (Göttingen: Vandenhoeck & Ruprecht, 2017), 22–25, 39–43.

Castagnino, but also many others).[16] The reasons for translating also varied widely, ranging from erudition (as in the case of Laurence Humphrey, studied here by Pérez Fernández, or specialists translating specialised works, like the physician Hayâtîzâde Mustafa Feyzî Efendi studied by Durmaz), individual curiosity, proselytising (like the Jesuits at the Mughal court who appear in the chapter by Cattaneo) or political engagement, to official commissions or pecuniary necessity (like both Morgan and Antoniutti). Women often translated as it was the only way for them to get into print. As Peter Burke points out, in early modern Europe 'the vast majority of translators engaged in this activity only once or twice during their lives'.[17]

An exception to this is provided by the case of interpreters, such as the dragomans who played an important role in the interaction between Europeans and Ottomans. In order to be less dependent on Ottoman interpreters (frequently 'renegades') several European states developed specific programmes to train their own interpreters, like that created in France by Colbert in the mid-seventeenth century. However, as discussed in more detail by Giancarlo Casale in this volume, dragomans had good reasons to be secretive about their trade, and typically resisted combining their roles as technical interpreters with that of translators for the public at large. As a result, in the Ottoman world, many of the most important early modern translators remained, rather than 'professionals', those who came to translation by accident. This seems to have been the case for most translators in the early modern period, who sometimes started from only a rudimentary knowledge of the source language.

Given this huge variety in the backgrounds of translators, and the large number of individuals concerned, research into translators as individuals can only be part of a collective enterprise that will be steadily enriched as more individuals who translated are identified and more is known about them. Increased knowledge about them will in turn throw more light on the circulation of information and ideas, translation choices and inter-cultural exchange.

An aspect of this knowledge highlighted here concerns the mobility of those who translated. How far was an individual's mobility a motivating factor in their engagement in translating activity? The link, mentioned above, between travel writing and translation would seem to reinforce the connection, but mobility could take various forms and be motivated by different circumstances. There is the obvious case of missionaries, seen in Angelo Cattaneo's chapter, who trans-

---

16 I discuss this at greater length in 'European translators in global context', *Bloomsbury Cultural History of Translation, Vol. 4 A Cultural History of Translation in the Construction of the Global World (16th-late 18th c.* CE*)*, ed. R. Clements (London: Bloomsbury, 2026), ch. 1.

17 Burke, *Cultural Translation*, 12.

lated religious works into the language of the culture whose members they were attempting to convert, but who also sometimes translated works from that culture into their native language, thus also helping the global spread of knowledge.[18] One can also cite the case of exiles like Laurence Humphrey, the Jewish families expelled from the Iberian peninsula (see the chapter by Durmaz) or the French Huguenots after 1685, who were particularly concerned to translate English works in favour of toleration into French.[19] Did mobility facilitate translation and were translators more mobile, and if so, how and why? Sometimes mobility gave rise to the desire to make available what the individual had discovered abroad to their fellow-countrymen and to open their minds to new ideas and knowledge. For others, it was the development of intellectual networks, facilitated by correspondence, which made it possible for a sedentary scholar to connect with distant places and translate their productions.

But socio-cultural contextualisation also means that attention needs to be paid not only to the individuals who translated but also, as is obvious from the chapters in the present volume, to the constraints under which they were working and to the others who played a role in determining translation activity. To quote Stockhorst again:

> Questions such as the following became relevant: which individuals (scholars, publishers, sovereigns, etc.) or institutions (academies, publishing houses, universities, etc.) decided that certain texts should be made accessible in their own language, and what were the underlying interests (academic, educational, political, commercial, etc.)? To which translational concepts and techniques did the translators refer? How were the respective texts received in the target culture?[20]

A translator is rarely completely free to choose which texts to translate, and even those individuals not limited by institutional or commercial constraints reflected the outlook, values and priorities of their own society, the cultural dynamics mentioned above. In addition, the autonomy and agency of the

---

18 Such as the Jesuit missionaries in the seventeenth century, already mentioned. More generally, see Angelo Cattaneo, *Tradurre il Mondo: Le Missioni, il portoghese, e nuovi spazi di lingue connesse nella prima età moderna* (Roma: Bulzoni Editore, 2022).

19 See several of the case studies in Thomson, Burrows and Dziembowski, eds, *Cultural Transfers: France and Britain in the long eighteenth century*; and A. Thomson, 'Pierre Des Maizeaux and the (Huguenot) business of translation in the early eighteenth century', in *Ideas Across Borders. Translating Visions of Authority and Civil Society in Europe c. 1600–1840*, ed. G. Mahlberg and T. Munck (London: Routledge, 2024), 83–98.

20 Stockhorst, *Cultural Transfer through Translation*, 23.

people who translated were often limited not only by the interests of the institutions to which they belonged but also by the commercial considerations dictated by publishers and booksellers, as seen in Thomson's study of Joseph Morgan. Historians of translation are therefore increasingly interested in the publishing history of translations and also the material aspects of printed translations, drawing on the work of book historians.[21] An aspect of the mobility of texts to which contributions in this volume pay particular attention concerns the material changes in a work when it is published – or in the case of manuscripts, recopied, as we see in the chapter by Durmaz – not only in a different language, but also in a different place, as shown by both Lista and Castagnino. The material form of a work also plays a role in how the text is received; a different frontispiece, layout, or format of a translation can also affect the meaning of the work and its effect on the reader.[22]

Translations and those who translated need therefore to be situated carefully and the more knowledge we have of their embeddedness, the more we shall understand the circulation of ideas and information. The place or places in which they carried out their translations and the dynamics of interaction between languages in those places therefore need to be taken into account. The so-called 'spatial turn' has been particularly influential in the study of the history of science,[23] but space, or rather place, also needs to be considered in the history of translation. Here it is useful to quote the definition given by J.E. Malpas:

> The idea of place encompasses both the idea of the social activities and institutions that are expressed in and through the structure of a particular place (and which can be seen as partially determinative of that place) and the idea of the physical objects and events in the world (along with the associated causal processes) that constrain, and are sometimes constrained by, those social activities and institutions.[24]

21 See for example Thomas Munck, *Conflict and Enlightenment. Print and Political Culture in Europe, 1635–1795* (Cambridge: Cambridge University Press, 2019), 180–224.

22 For the study of paratexts, see for example *Thresholds of Translation: Paratexts, Print, and Cultural Exchange in Early Modern Britain 1473–1660*, ed. Marie-Alice Belle and Brenda M. Hosington (Basingstoke: Palgrave Macmillan, 2018).

23 See in particular David N. Livingstone, *Putting Science in its Place: Geographies of Scientific Knowledge* (Chicago: University of Chicago Press, 2003).

24 J.E. Malpas, *Place and Experience: A Philosophical Topography* (Cambridge: Cambridge University Press, 1999), 35, quoted in Charles W.J. Withers, 'Place and the "Spatial Turn" in Geography and in History', *Journal of the History of Ideas* 70 (2009): 642.

Place plays a role in determining which works were translated, but also in the choices made by the translator, which are influenced not only by the target language but also by the intellectual, social and political climate of the place, extending beyond the obvious question of censorship.[25] As Livingstone puts it: 'Every social space has a range of possible, permissible, and intelligible utterances and actions: things that can be said, done, and understood'; and 'as ideas circulate, they undergo translation and transformation because people encounter representations differently in different circumstances'.[26] The essays in the present volume therefore look not only at the linguistic mobility of texts but also their geographical mobility; the specific places in which they were translated and published could change the value given to a work and the way it was both presented and read (see for example the chapter by Lista). As historians of science have pointed out, a concern with place needs also to take account of connectivity and mobility between places, or 'knowledge in transit' as highlighted by James Secord,[27] in which translation plays a predominant role. The places discussed here cover a wide geographical area, but with a central focus on the different shores of the Mediterranean, a region which provides particularly rich material for a study of the issues raised by this volume. Contacts with much more distant places are also studied, demonstrating that long distances did not prevent connectivity or mobility, as is particularly evident in Cattaneo's discussion of the activity of Portuguese Jesuits at the Mughal court in the late sixteenth and early seventeenth centuries. The specificities of the Ottoman empire also appear in several contributions, but as they also demonstrate clearly, the border between Western print culture and Ottoman manuscript culture was never stable or impermeable. Indeed, three contributions deal with this question directly, and from multiple directions: the case of an Ottoman doctor who translated and commented on printed European medical texts (Durmaz), of an English traveller who published in translation a Morisco manuscript (Thomson), and of a former Ottoman palace official who simultaneously translated *both* Ottoman texts for publication in Europe *and* European texts for circulation as Ottoman manuscripts (Tezcan). It is also, incidentally, interesting to note

25 See for example Charles W.J. Withers, *Placing the Enlightenment. Thinking Geographically about the Age of Reason* (Chicago: University of Chicago Press, 2007), 50–57.

26 Livingstone, *Putting Science in its Place*, 7, 11. Livingstone discusses in this work the manipulation of images, but he does not discuss the effect of textual translation on the circulation of scientific knowledge.

27 James A. Secord, 'Knowledge in Transit', *Isis* 95 (2004): 654–672; Withers, 'Place and the "Spatial Turn" in Geography and in History': 637–658.

the persistence within Europe of the manuscript circulation of some translations that would not have received official approval (Lista).

Examples of how the networks of individuals, and sometimes their own mobility, facilitated the mobility of texts are provided by the case studies of specific translations and translation activity in this volume. The order of presentation corresponds to increasing spatial mobility in relation to the central focus on the Mediterranean, beginning with the geographical mobility of a classical Greek scholar within Europe in the first contribution, moving to contacts between the different shores of the Mediterranean, and finally to the Portuguese Jesuits who travelled to the Mughal empire. Looking at very different cases of mobility and different types of language interaction (the subject of Giancarlo Casale's reflections), these contributions can only provide an initial and obviously partial picture of this vast domain.[28] We can, however, identify certain transversal themes which emerge, the first being obviously individual mobility.

We meet here a diverse group of mobile translators. Some were polyglot scholars as well as translators whose spatial mobility was linked to religious difference. Laurence Humphrey, whose importance is demonstrated by José María Pérez Fernández, was a Protestant scholar from Oxford, who fled England in 1555 and sought exile in Basel, where he worked for humanist printers before returning to Oxford on the accession of Elisabeth I in 1559. The chapter, which studies his editions of Greek manuscripts and translations into Latin, also brings out his participation in international networks of Greek and Hebrew scholars. The seventeenth-century 'renegade' Ali Beg/Ali Ufki/Albertus Bobovius, studied by Baki Tezcan, was also part of wide intellectual networks and seems to have had close connections with the Protestant world. Originally a European Christian from Poland or Western Ukraine, fluent in Italian and Latin and possibly trained by Jesuits early in life, he was subsequently 'Ottomanised' through an experience of enslavement, forced conversion, and training in the Ottoman palace. As Tezcan shows, he emerged at the far end of this experience still easily conversant with his European intellectual collaborators, yet deeply devoted to his Ottoman patron (and former master), and firmly rooted in the faith and culture of his adopted society, in which he could combine roles as a 'lover of Christ' (*hûb-Mesîhî*), a Muslim and an ecumenical humanist in a manner that was quintessentially Ottoman. Hayâtîzâde Mustafa Feyzî Efendi, the subject of the chapter by Tunahan Durmaz, also a convert to Islam, was

28 We regret in particular the absence of women translators, which in part reflects their more restricted mobility. A projected chapter on Maria Petrettini could unfortunately not be included as the author had to withdraw for personal reasons.

from a Jewish family expelled from Spain who had settled in Istanbul. A distinguished and highly respected physician, he was in contact with a network of Jewish doctors throughout Europe, which helped give him access to many contemporary European medical works; Durmaz shows how he translated them in his own work, supplementing these translations with observations from his own practice. These two cases of converts acting as translators are far from exceptional. In fact, for the entire period of Ottoman history up to the middle of the eighteenth century, there are but a handful of examples of translators from European languages into Turkish or Arabic who were *not* either European slaves, or Western renegades, or converts to Islam from one of the Ottoman Empire's local communities of Christians or Jews.

Annibale Antonini, one of the translators who features in Giovanni Lista's study of Italian translations of Fontenelle's *Entretiens sur la pluralité des mondes*, who was from a wealthy Southern Italian family, was also able to establish influential networks that stood him in good stead when he settled in Paris and made a name for himself as a teacher of Italian and propagator of Italian culture. As Lista shows, his version of Fontenelle's text, published in Paris, was part of his strategy for presenting himself as the promoter of the Italian language and had a very different emphasis from the translation produced by the scientifically oriented polymath Bernardino Vestrini in Cortona.

Unlike these figures, the obscure Joseph Morgan (the subject of the chapter by Ann Thomson) illustrates the difficulties encountered by an outsider, who apparently did not possess the social position or intellectual networks to succeed and prosper. We do not know where he was educated or why he spent twenty years in North Africa at the beginning of the eighteenth century; he wanted to make a name for himself as a historian and a writer, and his activity as a translator seems to have been in large part conducted for reasons of pecuniary necessity. At the same time, his translation of a Morisco manuscript on Islam was at least partially motivated by his wish to make that religion better known to his compatriots and to reduce their prejudices against Muslims. In this translation, as in his others, he added notes and comments to make his own position clear.

This was also the case of the Jesuit Antoniutti, who had otherwise little in common with Morgan and enjoyed a very different social status. It was his ecclesiastical career that led to his geographical mobility, social connections and intellectual network, as Alessia Castagnino shows. These travels – and in particular the time he spent in Venice, Vienna and Istanbul – encouraged his interest in translation. In addition, Castagnino brings out the importance of the fact that he finally settled in Venice, in view of the role played by the discussions there in the later eighteenth century on how to renew their governmental and economic system; this atmosphere encouraged translations of foreign works, particularly from the English, that represented new thinking and discussed reforms that could be adopted. He thus shared with Morgan a political motivation for his choice of works to translate.

The importance of place is particularly striking in the case of the Jesuit missionaries at the Mughal court in the 1580s, analysed here by Angelo Cattaneo. Rodolfo Acquaviva and his companions were encouraged by Akbar to study Persian and translate the Gospels, but their hope that he was interested in converting to Christianity was misplaced and they were massacred by the inhabitants of a village to which they were sent by their superiors. Their translation of not only gospels and catechisms but also classical and Christian authors is highlighted in this chapter, as well as the transformation in linguistic knowledge, awareness and practices that took place within the framework of the Catholic missions from the mid-sixteenth century onwards.

This example highlights the other main aspect of mobility that runs as a theme through these chapters, namely the mobility of knowledge and the circumstances favouring this mobility, including place. Pérez Fernández shows how Lawrence Humphrey's years in Basel brought him into contact with Greek manuscripts and allowed him to develop his practice of translation as a method for transferring this classical learning in conjunction with more recent works. Lista's discussion of the Italian translations of Fontenelle's dialogue brings out how, unlike Antonini's translation – conducted for a Parisian audience in the context of the defence of the Italian language – Vestrini's *Trattenimenti* reflects the scientific debates in the Academy of Cortona; his translation became a site for discussion of Newtonian and Cartesian theories, in the light of the Galilean tradition. And, as we have seen, Antoniutti's translations contributed to the transfer to Italy of recent thinking in economics, encouraged by the particular circumstances of intellectual debate in Venice. This was doubtless more successful than Joseph Morgan's attempt with his translation of Rabadan's poem to transfer to Britain more accurate information about Islam and its founder, which does not seem to have found a receptive audience in London. His historical work, however, whether in his translation of *The History of the Revolutions in Spain* or his own history of Algiers, was probably more successful in the transfer of knowledge by appealing to curiosity about these countries and their history.

Similarly, it was Ali Beg/Bobovius's presence in Istanbul, employment as a dragoman and contact with European diplomats that led to his role in the transfer of knowledge between cultures (Tezcan). In addition to an account of the Ottoman imperial palace, two Latin treatises on Islam, teaching material and a Turkish grammar and dictionary for European readers, he translated a Latin version of a Church of England catechism into Turkish as well as producing a Turkish translation of the Bible and of Comenius's *Janua linguarum reserata*. As his revised translation of the Bible circulated into the twentieth century, he succeeded in domesticating the Bible for a vernacular Turkish-speaking audience. In a different field, as Durmaz shows, Hayâtîzâde's *Resâilü'l-müşfiyye* contributed to knowledge transfer by providing translations of medical works in Latin by different classical authorities but also more recent European authors reflecting contemporary trends in medicine, combined with his own observations of disease. He thus contributed to the process of the vernacularisation of Latin medicine in the Ottoman world, and his work continued to circulate widely in manuscript form in both the Ottoman Empire and Europe.

Several of the other contributions, as we have seen, also bring out the interactions between manuscript and printed circulation, and the afterlife of works through translation. Taken in conjunction, therefore, the different chapters of this volume, dealing with a variety of individuals, languages and geographical and cultural spaces, highlight important questions from new and different angles. They also demonstrate the extent to which the study of translators linked to that of mobility can provide a deeper understanding of many aspects of the history of the early modern period.

CHAPTER 1

The Wages of Exile: Translation and Mobility in Laurence Humphrey's *Interpretatio Linguarum*

José María Pérez Fernández

A Return Ticket to Basel: Humphrey's Production as a Marian Exile

Born at Newport Pagnell, Buckinghamshire, around the third decade of the sixteenth century, Lawrence Humphrey started his career in academia as a pensioner at Christ College, Cambridge, in 1544. Three years later he moved to Oxford, where he became demy of Magdalen, the college to which he would be permanently attached for the rest of his life. In Oxford he held the positions of lecturer in natural and moral philosophy in 1552 and 1553 respectively, and like so many other Protestant figures he was forced to leave England for Continental Europe after the death of Edward VI and the coronation of Mary Tudor; this would turn out to be an enlightening and most fruitful period of exile. He arrived first in Zürich in 1554 and moved to Basel in 1555; he enrolled in the university and joined the international team of scholars who worked as editors, translators and correctors for the publishing workshops of humanist printers like Froben, Episcopus and Oporinus. As we shall see, he also collaborated with them as the author of paratexts for some of their major publishing projects. In fact, he continued to do so after his return to Oxford in 1559, where he took up his academic career first as Regius Professor of Theology, and then as President of Magdalen.

Humphrey's Basel was a hub for intellectual production and a refuge for Protestant exiles who found employment in its thriving publishing business. An international group of zealous reformers and well-educated polyglot scholars, they shared among them the skills they had already acquired in their respective countries, and once some of them returned home – as Humphrey did – they brought back the knowledge, the experience and the contacts they had gained abroad. Humphrey's production and that of other European exiles all constitute excellent cases of texts and knowledge in motion, which also included the circulation of the sorts of skills required for the editorial preparation and translation of ancient manuscripts and modern titles. Humphrey's production in Basel, which as we shall see consists in texts published roughly between 1557 and 1560, is both part and the result of the intricate paths trodden by

 | DOI:10.1163/9789004731806_003

manuscripts throughout networks that went from East to West across the Mediterranean. The manuscripts were then appropriated by patrons, collectors, scholars and publishers who repurposed and redistributed them for European readers in their original languages, alongside their Latin and vernacular translations. As this chapter will demonstrate, Humphrey's fruitful exile yielded a series of texts that amount to a quasi-encyclopaedic compendium of classical and contemporary disciplines, in terms not just of mere theoretical speculation, but also of their mobilisation within the intellectual, political and religious controversies of the age. Translation, understood as both hermeneutics and a method for the administration and communication of information and knowledge, is the common thread that runs throughout his work.

The texts Humphrey penned and published in Basel included a couple of titles that addressed two of the main intellectual debates brought about by the religious and political crises of these years. His *De religionis conservatione et reformatione vera* (Oporinus, 1559) was an essay on the Reformation, its doctrinal principles and their political ramifications, whereas *Optimates, siue de Nobilitate* (Oporinus, 1560) printed Humphrey's Latin translation of a short text by Philo of Alexandria on true nobility. This version of a Greek original was published as an appendix to Humphrey's own long treatise, which was itself an updated interpretation and copious re-elaboration of the moral and political philosophy of this Jewish philosopher of the first century AD. Dedicated to Queen Elizabeth and the gentlemen of the Middle Temple, *Optimates* embraces the tenets of traditional aristocracy as it also seeks to legitimise the incorporation of new men into the ranks of the regimes brought about in the mid sixteenth century by the downfall of well-established political and religious institutions.[1] These two texts were published in quick succession during the period that saw Humphrey's transition from Basel back to England, where he joined the Elizabethan establishment. Humphrey had in fact created close ties with other important Marian exiles who were laying the ground for the reestablishment of the Reformation in England after the death of Mary Tudor and the political turmoil that ensued until her half-sister Elizabeth eventually reached the throne. Among several other names who frequently crop up as dedicatees or as the object of praise in Humphrey's texts, these political and religious agents included John Bale, John Foxe (both of whom also collaborated with Oporinus), Francis Walsingham (who had wandered through

1 The full title is *Optimates, siue de Nobilitate, eivsque antiqva origine, natura, officijs, disciplina, & recta ac Christiana institutione Libri tres. Laurentio Hvmfredo autore. Adiunctus est proper utilitatem & affinitatem argumenti, Philo Iudaeus de Nobilitate, Graece & Latine: Edodem interprete. Basileae, per Ioanem Oporinum.*

cities like Padua and Basel during his years in exile) and Edwin Sandys (who had spent some time in prison under Mary Tudor before fleeing to Strasbourg and Zürich). His *Interpretatio* is dedicated to Sir Thomas Wroth, who had held important posts under Edward VI and had had difficulties (including exile in Italy and Strasbourg) under Mary I, until he returned to Parliament soon after Elizabeth's accession. This background explains Humphrey's timely publication in 1559 and 1560 of two works that address some of the main doctrinal and ideological pillars required by the new queen and her political agents to legitimise and cement the institutional foundations of her rule. Humphrey collaborated with the monarch and her religious settlement from a position of critical loyalty, and he comes across as a Reformer with Puritan inclinations who nevertheless strove to exercise a pragmatic and possibilist attitude vis à vis the authorities. This compromising stance is confirmed by his successful institutional career in Magdalen and his connections with some of the most devoted English Puritans.

Other than these two works, Humphrey also produced texts directly related to important disciplines like lexicography, translation and in general the different methods within the toolbox of classical and humanist philology, all of them employed for the interpretation, edition and publication of texts ancient and modern, most, if not all, with significant political, religious and cultural relevance. One of the earliest fruits of Humphrey's collaboration with Froben was his 'Epistola de Graecis litteris et Homeri lectione et imitatione', a preface to the monumental edition, in its original Greek, of the twelfth-century Eastern Roman scholar Eustathios of Thessalonike's commentaries on Homer (*Copiae cornu sive Oceanus enarrationum homericarum*, Basel: Frobenius, 1558), edited by one of Humphrey's contemporaries, Hadrianus Junius (i.e., Adriaen de Jonghe, 1511–1575). This text is revealing in many respects: Humphrey presents the book as a token of gratitude to the fellows of Magdalen College and describes himself as a modern-day Ulysses with Oxford as his Ithaca. His declared aim was to promote the study of Greek in his homeland, and to that effect he recounted the long history of the language, from its birthplace in Antiquity, its subsequent dissemination across the Mediterranean and then its transfer, thanks to the intervention of divine providence, on to England.[2]

2 After a series of rhetorical questions that list the temptations and trials encountered by Ulysses during his pilgrimage – all of which failed to make him forget Ithaca – Humphrey stresses how Cicero encouraged his readers to imitate the virtues found in Homer's poetry. Humphrey then embarks upon an emotional passage in which he says that, although he is in a very learned city like Basel and in the company of learned men, he misses England. This is followed by a long praise of Magdalen as a place of knowledge where many illustri-

This text thus rehearses the narrative used by sixteenth-century humanism and its political patrons to account for the long history of *translatio imperii* as a process for the legitimate appropriation of the cultural and religious capital of the Ancient World, a process which also involved methods for the interpretation of Scripture. Combined skills in Hebrew and Greek were part of the process, hence Humphrey's interest in a Greek-speaking Jewish Neoplatonist philosopher like Philo, perceived as one of the intellectuals responsible for the incorporation of Greek language, culture and philosophy into the doctrinal body of texts that sustained the main tenets of early Christian doctrine. This naturally includes a brief account of the arrival and distribution of Greek manuscripts, their transcription, editorial interpretation and eventual publication in print. Humphrey underlines the position of Greek at the root of the liberal arts, its role as the source of all knowledge and the foundation of the hierarchy of the disciplines.[3] This is why, he claims, a proper hermeneutic approach to the Homeric poems is fundamental, since they amount to a secular Scripture of sorts: the different episodes in the Odyssey and its protagonist are mirrors of human life that cast light on our condition as pilgrims.[4] Humphrey extols Homer as the universal poet, whose work comprehends all disciplines, including rhetoric, oratory and cosmography. The poem's proverbs and sayings, he concludes, are a source of general wisdom. But before reaching this sort of transcendental knowledge, the reader needs to master the skills that will allow for the proper interpretation of the poems, and these tools are facilitated by grammarians, rhetoricians and editors like Eustathios of Thessalonike and Hadrianus Junius, who will open up the obscure meaning of ancient Greek vocabulary.[5]

If this work on the philological nuts and bolts of Homeric hermeneutics evinces Humphrey's connections with international networks of Hellenist scholars, another of his paratexts confirms his contacts with some of the most important Hebraists of his age. Just one year before the Homeric treatise, in 1557, Froben and Episcopus published a monumental work in Hebrew lexicography, the *Dictionarium Hebraicum novum*, for whose paratexts Humphrey

---

ous men have been educated, until he returns to the comparison with Ulysses, a reference to William of Waynflete, the founder of Magdalen College, and an emphatic description of the edition he is presenting, as a gift of gratitude to the college ('Epistola de Graecis litteris et Homeri lectione et imitatione', in *Copiae cornu sive Oceanus enarrationum homericarum*, Basel: Frobenius, 1558, *2ʳ–*2ᵛ).

3 Further details in *2ᵛ–*3ᵛ.

4 'Continet tamen humanae uitae speculum. Nihil est eam nostra uita quam Odyssaea, id est Vlyssea peregrinatio', [*4ᵛ].

5 'Grammatici uerborum significationes proprias & antiquissimas hauriant', [*4ᵛ].

composed a Neolatin poem. The author of the dictionary was Johannes Forster (also spelled Furster, 1496–1558) a close acquaintance of Luther, whom he had aided in his translation of Scripture. Forster was also connected with the Hellenist Philip Melanchthon, and had learned Hebrew under Johannes Reuchlin, who also penned a brief paratext for the *Dictionarium*. Forster's own introduction is addressed to Ulrich Fugger III (1528–1584), a dedicatory preface which, like Humphrey's introduction to the *Oceanus erranationum homericarum*, constitutes another textbook case of *translatio imperii*, the patrons and networks through which ancient texts circulated and eventually reached European centres of political and cultural power. Ulrich Fugger was a member of the well-known banking family, and was not only a keen collector of Latin, Greek and Hebrew manuscripts but also a Hellenist himself, the patron of printers and publishers like Henri Etienne and the facilitator of different editions of classical Greek texts which went into print for the first time thanks to his intervention. Forster compares the Fugger family in the sixteenth-century German-speaking world to the role of the Medici in the Italian Peninsula in the previous century: they were both patrons of learning, book collectors and the creators of famous libraries. Both were, in short, wealthy banking families who facilitated the acquisition and transference of important manuscripts from the Greek-speaking territories of the Eastern Roman Empire to Europe. Once there, a host of scholars, publishers and printers would transcribe, edit, translate and eventually publish them in a diversity of formats for different sorts of readers.

Alongside Homeric poetry and Hebrew Scripture, there is a third component to Humphrey's participation in international projects that furnishes another important pillar in the secular and religious foundations of Reformed Europe, closely connected with his interest in the work of Philo of Alexandria. This is his translation from Greek of three dialogues by Origen of Alexandria (c. 185–c. 253 AD), the *Contra Marcionistas Dialogi nunc primum Latio donate Laurentio Humfrido Anglo interprete*, which appeared in Oporinus's edition of this prolific theologian, two hefty tomes first published in 1557 with a long introduction by Erasmus on Origen's life and work. A collaborative project which required the intervention of a large team of scholars and translators from all over Europe, this edition in folio format ran to almost 2,000 pages and confirms Humphrey's international reputation as a Hellenist and translator.

Humphrey would muster all his international contacts, interdisciplinary familiarity with classical rhetoric and poetry, close acquaintance with the contemporary humanist canon and expertise in Greek and Hebrew for the *Interpretatio Linguarum*. In his attempt to bring home to Oxford the religious and philosophical capital of the Ancient World, to transfer the wisdom and doctrines formulated in Pagan Greek epic poetry, in the Hebrew and Greek Scriptures, in the works of early Christian theologians like Origen and in the moral and political philosophy of Philo of Alexandria, Humphrey engaged in his own project of *translatio imperii* from all these different sources to Elizabethan England.

The *Interpretatio linguarum* and the other texts he produced in Basel address most, if not all, the theories and practices that concern the production and transfer of knowledge with translation at their centre. The rest of my chapter will provide a necessarily brief survey of some of the most important. They include the relation between translation and the reform of the disciplines, a question in which Humphrey evinces his close reading of Juan Luis Vives's work. This topic is also closely connected with the ways in which the doctrine and practice of translation relate to grammar, rhetoric and lexicography, and more specifically to their practical application to the different techniques involved in the comparative philological analysis of original versions in manuscript in order to produce critical editions and establish a *clean* text. The ideal of the clean text as the result of a systematic process of philological archaeology was in turn a prerequisite for the next step in the task of opening up an ancient work for new readers through translation; this involved an approach to translation as hermeneutics and to the translator as an über-reader of sorts who was fully equipped to facilitate knowledge of a higher order, both sacred and profane. Finally, Humphrey's texts demonstrate the importance that he accorded to the development of methods and the publication of materials for teaching and learning foreign languages. In short, all the doctrines he absorbed and synthesised, alongside the practical know-how which he harnessed for the composition of his own texts – in particular, as we shall see, the comprehensive and quasi-encyclopaedic *Interpretatio* – are the result of a mind that was fully conversant with the different threads that wove together the intellectual Europe of his age.

Recent scholarship by Sheldon Brammal on Humphrey's *Interpretatio* has stressed its links to Joachim Périon's *De optimo genere interpretandi* (1540), and more specifically to the way in which that French scholar formulated his ideas on translation through the tenets and language of the Ciceronian controversy. Cicero is indeed a constant presence in Humphrey's treatise and Périon is also quoted in certain passages.[6] However, a close reading of the *Interpret-*

6 Sheldon Brammal, 'Laurence Humphrey, Gabriel Harvey, and the Place of Personality in Renaissance Translation Theory', *The Review of English Studies* 69 no. 288 (2017): 56–75. Other than Brammal, scholars who have studied Humphrey include Eleanor K. Merchant, '"Doctissimus pater pastorum": Laurence Humphrey and Reformed Humanist Education in Mid-Tudor England' (PhD diss. Queen Mary, University of London, 2013, https://core.ac.uk/

atio confirms that the entire pantheon of classical thinking on language and translation, from Aristotle to Cicero, illustrated with the texts of the Greek classics from Homer to the great tragedians, as well as contemporary humanism – Valla, Erasmus, Vives, among several others – all form part of Humphrey's rich and complex mosaic.

Humphrey's *Interpretatio Linguarum* and Its Genealogy

The *Interpretatio* is in fact not just the most important and original outcome of Humphrey's productive exile in Basel, but a compendium of classical, late medieval and contemporary doctrines concerning translation. It includes both theory and a host of fundamental ancillary disciplines inextricably involved in its practice, such as rhetoric, grammar, lexicography, morphology, syntax, and in general the array of philological methods required for the textual administration and engineering of information and knowledge recorded in script. His work incorporates the linguistic thought of Lorenzo Valla, a scholar whom Humphrey also admired as translator of Herodotus and Thucydides and as the author of a comparative textual analysis of the Latin version of the New Testament in the Vulgate with several original Greek manuscripts, the *Collatio Novi Testamenti* (1447). The latter title in turn profoundly influenced Erasmus's own edition of the Greek New Testament, the *Novum Instrumentum* (published in 1516 by Froben in Basel), a work which Humphrey also held in the highest regard. The intellectual framework of Humphrey's *Interpretatio* also includes the doctrines of Juan Luis Vives on language, his proposal for a reform of the disciplines and the general principles of his pedagogy. Beyond these, one can find the usual names from the classical and post-classical world: Aristotle (above all his *Rhetoric*), Cicero and Jerome, among several others.

The close functional connections between hermeneutics and translation on the one hand, and their practical application to religion and politics on the other, are formulated in Humphrey's dedicatory preface to Sir Thomas Wroth.

download/77038766.pdf, accessed on 06/02/2025), in particular chapter 2. There are also references to Humphrey in Glyn P. Norton's *The Ideology and Language of Translation in Renaissance France and Their Humanist Antecedents* (Geneva: Librairie Droz, 1984), and Frederick M. Rener's *Interpretatio: Language and Translation from Cicero to Tytler* (Amsterdam and Atlanta: Rodopi, 1989). Neil Rhodes translates into English a series of relevant passages from Humphrey's *Interpretatio Linguarum* in Neil Rhodes, Gordon Kendal and Louis Wilson, eds. *English Renaissance Translation Theory* (London: MHRA, 2013), an anthology of texts which is extremely helpful for the contextualisation of Humphrey's eminently international background within his more immediate domestic context.

In its opening paragraph Humphrey establishes a major division of knowledge into two main fields: religion, which he couples with the discipline of theology, and literature, which he pairs with philosophy and philology.[7] Humphrey also extols the job of the translator, whose function in the field of religion is to establish a textual conversation with God so he can first interpret and then open up Scripture for the community of believers, a process that grants the translator the sublime privilege of giving God himself a voice. 'What can be more sublime than acting as God's interpreter?', he asks in a dramatic crescendo, 'than establishing a conversation with God? Nay, than giving God himself a voice?'[8] This was a widespread idea among Lutheran translators of Scripture, also echoed in the title page of Francisco de Encinas's first Castilian version of the New Testament (Antwerp, 1543), which proudly proclaims that his text is God speaking ('Habla Dios'). Like Emmanuel Chrysoloras in the late fourteenth century and Lorenzo Valla in the fifteenth, Humphrey and his fellow Lutheran translators were devoted to finding ways in which the 'veritas' or 'proprietas hebraica' and 'graeca' of the Old and New Testament could be brought to light and reformulated in vernacular translations that were the result of a long and complex process of textual archaeology conducted with the tools of grammar and rhetoric.[9]

In the first part of his second book (*De ratione imitandi*) Humphrey elaborates further on this subject by invoking Greek models alongside Cicero while discussing the important role of translation and imitation in the discursive articulation of the disciplines and the establishment of their hierarchy, as well as the creation of the different textual genres and linguistic registers that codify them (e.g., historiographical prose, poetry, or prose fiction). Other than its rel-

---

7 See a2^{r}

8 'Quid sublimius quam Dei interpretem agere? Quid diuinius quam cum Deo colloqui? imo quam Deum loquentem facere?' (a3^{r}).

9 Manuel Chrysoloras (Constantinople, c. 1355–Constance, 1415) was an Eastern Roman diplomat who brought Greek texts and skills in that language first to Florence and then to Milan and Rome at the turn of the fourteenth century. On the group of Hellenists who helped to bring original Greek manuscripts to Western Europe from the receding territories of the Eastern Roman empire see N.G. Wilson's classic *Scholars of Byzantium*, (London 1983). Further details on Chrysoloras can be found in Marianne Pade's 'Chrysoloras on Translation: a note on the meaning of *proprietas graeca*', in *God Latin. Festskrift til Peter Zeeberg i anledning af hans 60–årsdag den 21. april 2017/Studies in Honour of Peter Zeeberg on the Occasion of his Sixtieth Birthday 21 April 2017*, ed. Birgitte Bøggild Johannsen, Karin Kryger & Karen Skovgaard-Petersen. Redaktionelmedarbejder: Marie Louise Blyme, *Renæssanceforum* 12 (2017): 53–60. See also James Hankins, 'Chrysoloras and the Greek Studies of Leonardo Bruni', in *Crisolora e il ritorno del greco in occidente. Atti del Convegno Internazionale. Napoli, 26–29 giugno 1997*, ed. Riccardo Maisano & Antonio Rollo (Naples: Istituto Universitario Orientale, 2002), 175–203.

evance for the interpretation and translation of the Bible, Humphrey is aware of the importance of translation for establishing the terms in which the discourse of a particular discipline is formulated, hence the need for accuracy in both semantic and pragmatic terms. This is also the reason why a good translator needs to be well versed in grammar and all its different subdisciplines.

There is plenty of evidence to confirm Humphrey's systematic approach to his subject. For example, in the opening pages of his second book there is a section on the aims of the translator ('De imitatione, exercitatione et fine interpretis', 209–215) that introduces different opinions regarding imitation and interpretation; it includes aspects of the Ciceronian controversy with explicit references to the debates between representatives of each side, such as Erasmus versus Longolius, Politian versus Cortesi and Bembo versus Pico. These were controversies which also concerned the translation of the Gospels, since there were those who would render Christ and his apostles preaching in the sort of Latin that the Roman rhetorician would use. This reproach was not dissimilar to the objections, in the previous century, to Leonardo Bruni's attempt to translate Aristotle into Ciceronian Latin.[10]

Humphrey's decidedly normative purpose is visible in the second part of his first book (*Secunda pars, interpretis officium*), where he lists the four main features, or prerequisites, of a good interpreter / translator, and then devotes several pages to describing and exemplifying each of them: *natura idonea* (116), *doctrina elegans* (125), *religiosa fides* (170) and *singularis diligentia* (180). A diagram and several indexes that close his first book – whose overall title is *De ratione interpretandi* (1–208) – give us another revealing measure of Humphrey's comprehensive scope. One of them lists in alphabetical order all the authors he has quoted in the preceding pages, all thirty-three of them, from Augustine to Xenophon (204–205). These include fifteenth-century scholars like Valla, alongside Humphrey's near-contemporaries, such as Erasmus, Vives, Longolius, and the publisher-scholar Aldus Manutius; the classical authors range from Plato and Aristotle to Cicero and Origen as well as the translators (or *interpreters*) of the Septuagint. But Humphrey's *Interpretatio* does not just nod in the direction of the entire canon of authors on grammar, rhetoric, the interpretation of Scripture, and how they all relate to translation theory and practice: he also incorporates his own compatriots into this tall and broad family tree. The final pages of the *Interpretatio* (526 ff.) feature his praise of Oxford

10 On this debate see T. González Rolán, et al., *Humanismo y teoría de la traducción en España e Italia en la primera mitad del siglo XV. Edición y estudio de la* Controversia Alphonsiana: *Alfonso de Cartagena vs. L. Bruni y P. Candido Decembrio* (Ediciones Clásicas: Madrid, 2000).

and Cambridge scholars with a list of the best recent and contemporary English poets, translators, lexicographers and grammarians: Geoffrey Chaucer, the Earl of Surrey (1517–1547), and Sir Thomas Wyatt (1503–1542) in literature and poetic translation, alongside the learned Hellenist and devoted Protestant John Cheke (1514–1577), or lexicographers like Thomas Cooper (c. 1517–1594), a Magdalen fellow who at the behest of the printer Thomas Berthelet curated in 1548 a new and revised edition of Sir Thomas Elyot's 1538 Latin-English dictionary.

The main topics covered in the early sections of the *Interpretatio* are also listed towards the end of Book One in a diagram that precedes the author index and amounts to an anatomy of translation. Here Humphrey provides basic notions (*Interpretatio definitio*) for his readers, a detailed classification of different types of translation (*Interpretandi genera*), as well as normative approaches to it (*Praecepta seruanda*), which include principles explicitly drawn from Aristotle (e.g., *Differentia uerborum, sensu, sono, ui, ex Arist.*). Another section of this index is titled 'Verborum in hoc libro explicatorum synopsis', which lists a series of Greek words and concepts in Latin to whose meaning, interpretation, and translation Humphrey devotes particular attention, and which he uses to illustrate and prove his arguments. For example, he directs his readers to a section which Humphrey indexes as 'Bonus, bene, Graec.Lat.Heb.Anglice' (206), in which he lists and briefly surveys the different meanings of the noun *bonum* and its correlates as an adjective and adverb, with examples of their usage in different contexts, authors (e.g. Cicero, Aristotle), and languages (Greek, Latin, Hebrew, English), with a profusion of commentary to demonstrate in what manner in each of these cases these morphological items should be rendered so that their translation can be as accurate as possible.[11] This is, in other words, a technically sophisticated disquisition which confirms the functional links between translation practice, translation theory and subdisciplines like lexicography and morphology. It also exemplifies the comprehensive way in which Humphrey seeks to couple, with a pragmatically pedagogic intent, the general principles of grammar and translation theory and their practical application to particular cases. This attitude pervades all the different sections in the *Interpretatio*, and they are encapsulated in an expression used by Humphrey towards its end: 'Non satis est hic scire, sed agere' (494), which can be freely translated as 'it is not enough to know this: one must also learn how to do it'.

What emanates from these and other sections – such as the first part of his first book – is therefore a systematic and nuanced approach to translation which seeks to harmonise and justify in very pragmatic terms differ-

11 More details in 104ff.

ent approaches or solutions to different sorts of specific problems. This also includes Humphrey's exposition of the main virtues of a good translation, which he indexes as *plenitudo, proprietas, puritas,* and *aptitudo* (201). With *proprietas*, for example, a translation seeks fidelity and accuracy, which leads Humphrey to suggest that in some cases the preservation of the *proprietas* of the original text requires the reproduction of the original word in the target language.[12] This principle had been already formulated in his first book, where Humphrey stressed the need to be faithful and never deviate from the original through amplification or diminution of its sense and significance. He resorts once more to the authority of Cicero, who defended the use of Greek words in Latin translation to preserve their *proprietas*, an argument which is reinforced by a list of several examples of Greek vocabulary borrowed by Latin in order to preserve this virtue.[13] The implementation of *proprietas* also requires an understanding of the *natura* and *vis* of vocabulary in the original, which as we shall see amounts to an accurate understanding of its semantic content alongside its illocutionary and perlocutionary potential.[14] These are also the same requirements for a *translatio plena*, i.e. one that has previously grasped not just the mere meaning of the original but also the intentions in the mind of the author when the text was originally articulated. Only then can the sense be rendered integrally (*sensum reddit integrum*) without substracting or adding

12 Humphrey formulates *proprietas* as a sort of authentic but nevertheless quasi intangible quintessence in each of the languages involved, such as *Latinitas* among Romans ('Romani Latinitatem, Hellenismum Graeci, Atticismum Attici, Iudaei Hebraismum'), which materialises in linguistic terms the *truth*, the *propriety*, and the *fidelity* inherent to each of them: 'uere, proprie, fideliter', three adverbs whose equivalent adjectives and nouns systematically recur throughout the *Interpretatio* 'Sequitur, ut de proprietate dicendum esse uideatur: ea *Idiotismus dicitur, qui in omnibus linguis conspicitur*. Romani Latinitatem, Hellenismum Graeci, Atticismum Attici, Iudaei Hebraismum uocant, *quum scilicet quaeque res suo nomine signatur uere, proprie & fideliter*.' (41, further details in 41–47; italics are mine).

13 Elsewhere (see 42) he keeps insisting on Cicero and his instructions to the effect that it is sometimes necessary, for the sake of accuracy, to use the original Greek words in Latin when an accurate equivalent cannot be found, which is very close to the debate between Bruni and Alonso de Cartagena, as well as the doctrines of Manuel Chrysoloras, which evolved into the Ciceronian debate in the early sixteenth century. This admonition is followed by a section (51–52) on the translation of Scripture and the use of Ciceronian Latin, with references to Longolius, and the idea that, although he does accept that the word of God ('verbum diuinum') can indeed express itself in Latin translation with Ciceronian elegance and purity ('eleganter, pure, ornate'), Christ in the Gospels must be made to speak in his own voice ('ut Christus maneat Christus, Euangelium Euangelium', see 52, and further details in the following pages).

14 'Maxime uero naturam uimque uerborum cognitam habere oportet & intellectam' (44).

anything (*mentem autoris ad plenum omninoque consequatur, in qua nec absit quicquam nec superet*).[15] This section is followed by a long passage in which Humphrey anatomises the components that make up an *interpretatio plena*, and as he does so he establishes further sub-classifications that deconstruct the sentence and discourse in general (*oratione*) into its different components (e.g. *res, sententiae, forma, figura, numerus, collocatio, conformatio*).[16] This is a complex passage whose sources can be traced back to Aristotle as well as other Greek rhetoricians and grammarians who stand behind Cicero's doctrines. As he usually does, Humphrey preaches by example, since some of the technical rhetorical terms he uses in these pages are quoted in their Greek original for the sake of accuracy. One of the constant features of Humphrey's text is that he insists on the same, or similar, idea(s) in a recurrent manner, repeating the same principles with great *copia* combined with variation in the terms he uses, and always interspersing his doctrine with a profusion of examples and practical cases. And so, soon after the previous passages, he keeps insisting on the importance of taking into consideration the nature (now referred to as *forma*) and the *tenor* of the text.[17] The cases he uses to illustrate all these principles are taken from Scripture (both Greek and Hebrew), as well as secular works, including essays in political philosophy (e.g., Aristotle's *Politics*), and from other Greek texts rendered into Latin by Lorenzo Valla, whom he sets up as an example of an accurate translator. Humphrey concludes this section by warning that although there may be no such thing as a perfect synonym, great care and detailed study of the most important authors is of paramount importance, so that the translator can acquire a diverse and copious lexicographical hoard in both languages involved. This is fundamental to enable him to produce an accurate and nuanced translation as close to the original as possible.[18]

---

15 '*Plenam uoco eam interpretationem quae sensum reddit integrum*, ut sententia respondeat ex aduerso sententiae, & *mentem autoris* ad plenum omninoque consequatur, *in qua nec absit quicquam nec superet*.' (31–32, italics are mine).

16 '… tres esse res in omni oratione perlustrandas … primum *res* atque *sententiae*, tum *forma* & *figura*: post *numerus*, id est, ut ait Cicero, ipsa *collocatio conformatioque uerborum* …' (32, italics are mine).

17 See also 32–33, which are followed by a series of admonitions that elaborate on the general principle that the good translator must not just respect the intentions of the original author but also keep a reasonable via media.

18 As he usually does when describing the virtues of translation, Humphrey uses language which resonates with the doctrines of Chrysoloras and Valla: 'Interpretationis eius, quam quaerimus, uirtutes sunt hae, ut *plena* sit, ut *propria*, ut *pura*, ut *apta*. His enim fere partibus contineri statuo.' (31, italics are mine).

With the concept of *plena interpretatio* Humphrey proposes a rigorous approach that can faithfully render the original without losing anything and without adding anything spurious; while *puritas* postulates elegance in style and a clean expression, which Humphrey strives to formulate in very elegant Latin, once more preaching by example.[19] Humphrey elaborates on the virtues of *puritas* and *elegantia* with *copia* and above all with gusto. He certainly appears to feel very comfortable dealing with it in great detail, flaunting his own expertise as he does so, and indeed aiming to turn his own prose into an example of stylistic *puritas* and *elegantia*. He is avowedly inspired by Erasmus, whose translation of the Greek New Testament he proposes as a perfect model of *puritas*. This version, claims Humphrey, shuns barbaric Latin and renders Christ's own words in a pristine style which is the quintessential expression of Latinity (*clarius, Latinius, purius*).[20]

In many passages the *Interpretatio* does read like a handbook on style, which is of a piece with the well established humanist postulate – represented by the doctrines of first Valla and then Erasmus – proposing the creation of a plain yet elegant, sophisticated and versatile linguistic *koine* as the best vehicle for the communication of knowledge of all sorts. This would contribute to the creation of a community of speakers and authors who would employ this authoritative register to produce both original texts and updated translations which would make the religious, political and cultural capital of their originals available to contemporary readers. In Humphrey this common register results from a combination of Ciceronian and classical authorities with religious and doctrinal principles, which he expounds through his detailed exploration of all the nooks and corners, nuances and challenges posed by the practice of interpretation and translation. Alongside the great *copia* of samples he presents, this is what distinguishes the *Interpretatio* from other treatises on translation. For Humphrey tends to be more comprehensive, systematic and at the same time more detailed than others who frequently had a more specific agenda in mind,

19 This section on *puritas* (*Tertia uirtus puritas*, 57–114) is much longer than the two that precede it. Here Humphrey discusses questions of style under the guise of Latin and Greek abstract nouns such as *puritas*, *καθαρότης* and *perspicuitas* (see for example 57–58, and 114–115). Although the opening pages do not have as many examples as their counterparts in the other sections, there is a very lengthy essay after this introduction, with many different references to Scripture as well as quotations from the classical Greek and Latin canons.

20 'Nec sane possum non laudare Erasmi nunquam satis laudatam diligentiam, dum reclamantibus multis, corruptissimo & exulceratissimo seculo, Nouum Testamentum ab illoto sermone ac barbarie primus uindicare, & pristino suo ac germano nitori restituere coeperit: ita ut Christus nobiscum clarius, Latinius, purius loqueretur.' (62).

such as Bruni, whose work was prompted by the controversies on the translation of Aristotle's moral and political philosophy, or Manetti, who mainly dealt with the different translations of the Hebrew Psalms. Humphrey pays close attention to the nuts and bolts as he also calls upon the more comprehensive linguistic, rhetorical and epistemological principles of Lorenzo Valla and Juan Luis Vives alongside the doctrines and example of Erasmus, all of whom provide both practical models and also a solid epistemological foundation for his work. All of this furnishes Humphrey's *Interpretatio* with an authoritative theoretical framework that structures and sustains the constant flow of particular examples that he uses to illustrate his arguments.

When discussing the qualities required for a good translator, Humphrey also describes how even different personal temperaments apply to the translation of different genres. The personal disposition one should expect from a good translator combines honesty with piety and diligence with faith; this gives Humphrey a chance to take a swipe at the Catholic church by accusing its authors, and above all its translators, of corrupting language and hence Christian doctrine, either deliberately or out of sheer ignorance. This is a type of ignorance which he describes as both barbaric and impious, for which the only remedy is 'true solid and sincere erudition' ('uera solida atque syncera eruditio'). This rests, in turn, upon a combination of knowledge and language: 'γνωσις dicitur, altera λόγου, cognitio & oratio'.[21] In fact, he uses this dichotomy between *γνωσις – cognitio* and *λόγου – oratio* (later on also rephrased in terms of *scientia* versus *sermone*) as the starting point for a series of thoughts on tools for the codification and communication of knowledge through the practice of interpretation and translation. These tools also involve the techniques and skills for the production and use of commentaries, notes and other paratexts, again following in the wake of the philological and textual models of Valla and Erasmus or the example of Eustathios of Thessalonike's commentaries on Homer in Hadrianus Junius's edition, all of which Humphrey knew very well.[22]

21 'Peruersae sunt artes quae uel linguam corrumpunt uel mentem: & indocta doctrina & illiberalis quae corrupti sermonis magistra est uel impietatis ... Fugiendae ergo tales ineptiae, unde cum ratione insanire discas, aut barbarus euadas aut impius: & inscientia eiecta, ac praua doctrina, succedat uera solida atque syncera eruditio Ea in duas est distracta partes: quarum una *γνωσις dicitur, altera λόγου, cognitio & oratio*. Quam partitionem ex D. Paulo desumpsimus.' (126–127, italics are mine).

22 Humphrey stresses, with his usual *copia* of synonyms or near-synonyms, the importance of this sort of critical apparatus, which must be read closely and attentively in order to grasp the true meaning of the text in as much detail as possible: 'Sed & interpretum *enarrationes*, siue *expositiones*, siue *Homiliae*, siue *sermones*, siue *Commentarij*, siue *Annotationes*, siue *Scholia*, siue aliae piae *tractationes* nostro interpreti diligenter sunt legendae

Grammar is also described here as the epistemological foundation for a general programme of studies that can familiarise the translator with the discipline or disciplines he must tackle, in which ideally he should be an expert too.[23]

There are in fact two ways of looking at Humphrey's *Interpretatio*: one consists in studying the long list of authorities he invokes, and the *copia* of ideas and quotations from ancient and modern authors that he uses to formulate and illustrate his doctrine and rules. The other approach looks into the way in which this synthesis of past and contemporary translation theory and practice prefigures, with the conceptual tools and vocabulary of classical and humanist grammar and rhetoric, what today we would call pragmatics and speech act theory. In other words, the *Interpretatio* can be used as a fulcrum to survey the path trodden by classical grammar and rhetoric, and by the archaeological and editorial tools of humanist philology, from their distant roots in the Hellenistic period to their reformulation in sixteenth-century Europe. This is followed by their eventual reinterpretation by twentieth and twenty-first century linguistics and pragmatics, the latter of which furnishes us with speech act theory as an instrument to account for these traditional doctrines on translation. These different views all deal with hermeneutics alongside the epistemological and social dimensions of language as the repository of knowledge and the creator of human communities, as formulated by influential sixteenth-century authors like Juan Luis Vives in his *De disciplinis*.[24]

In page 130 of the *Interpretatio* Humphrey once more proclaims the role of grammar as one of the foundations for a general programme of studies for the education of a proper translator who engages in a hermeneutic approach to the original text as a necessary precondition for its subsequent rendering into the target language.[25] Here grammar provides the epistemological foundation

& attente. Habentur in illis non pauca, quae in usum suum conuertat, sed cum iudicio tamen excerpat consentanea' (129, italics are mine).

23 See 129–130. A few paragraphs later Humphrey stresses once more the importance of language learning, with an emphasis on Greek and Hebrew, for the translation of Scripture as well as works from the classical canon (see 134 ff.).

24 According to Vives, language 'is the treasury of culture and the instrument of human society' (*Ioannis Lodouici Viuis Valentini De disciplinis libri XX* (Antwerp, 1532), fol. 97r. English translation by F. Watson in Juan Luis Vives, *On Education. A translation of* De tradendis disciplinis, with an introduction by Foster Watson and a foreword by Francesco Cordasco (Totowa, NJ: Rowman & Littlefield, 1971), 91).

25 On the Hellenistic tradition of grammar and textual criticism see Irvine Wilson, *Making of Textual Culture.* Grammatica *and Literary Theory 350–1100* (Cambridge: Cambridge University Press, 1994) and Francesca Schironi, *The Best of the Grammarians. Aristarchus of Samothrace on the* Iliad (Ann Arbor: University of Michigan Press, 2018). For a brief and excellent introduction to the role of grammar in the Middle Ages and its evolution

and the tools required for the techniques of textual engineering we have come to call humanist philology, which, in the words of Anthony Grafton, 'rethought the nature of reading itself'.[26] The epistemological value of grammar and its practical application through humanist philology has been stressed by scholars like Donald R. Kelley, who claimed that 'modern historical thought ... owes its very existence to the encounters between philology and other branches of knowledge. Indeed, *philology stands in somewhat the same relation to historical scholarship as mathematics stands to natural science*'.[27] With the combined aid of grammar and lexicography, Humphrey suggests a two-step approach to translation as interpretation. This consists, as we have already seen, first in grasping the sense or the semantic content, which he calls *rerum ratio* and *sensus* (what modern pragmatics would call locutionary force, or locutionary act). The next step seeks to understand the intention of the utterance as a whole (*tota vis sententiae*) within its original context, which roughly coincides with what pragmatics calls speech acts, and more specifically with what their illocutionary and perlocutionary functions originally intend and do eventually achieve upon hearers or readers.[28]

Let me briefly single out three passages that illustrate Humphrey's awareness of this problem and the necessity for a translator to know in as much detail as possible all the nuances in the meaning and connotations of synonyms and near-synonyms in the original text, whose contextual significance therefore must be properly interpreted before searching for an accurate equivalent in the target language. He does so when discussing different nouns in Greek that denote the group of individuals who make up a political community. He distinguishes between those which can have a derogatory value, such as *vulgus, plebs* and *turbam,* in contrast to others which are more technically neutral, as it were, such as *populum*. And in order to exemplify a correct translation of one

---

towards its place in sixteenth-century humanism see Brian Cummings, 'Grammatical Culture: Medieval to Renaissance', in *The Literary Culture of the Reformation: Grammar and Grace* (Oxford: Oxford University Press, 2007) 20–26.

26 'Humanist philology, in short, eminently deserves the attention that it has received from modern classicists and historians. Its proponents not only filled in the gaps and cleaned off the stains of the classical corpus, but rethought the nature of reading itself', Antony Grafton, 'Renaissance Readers and Ancient Texts: Comments on some Commentaries', *Renaissance Quarterly* 38:4 (1985): 615–649 (627).

27 *Foundations of Modern Historical Scholarship. Language, Law, and History in the French Renaissance* (New York and London: Columbia University Press, 1970), 9, italics are mine.

28 'Quoniam igitur, autore Cicerone, omnis aut ex reb. aut ex uerbis constat oratio: *primum ac praecipue rerum ipsarum habenda est ratio*, ut sensus eruatur, & *tota uis sententiae interpretatio comprehendatur*. Nam hic si claudices, peccatum est grauissimum' (418; italics are mine).

of these terms ('λαὸυ non plebem, sed populum'), he invokes Lorenzo Valla's translations of Greek texts.[29] This is followed by another passage on the different nouns used in Greek and Hebrew to denote 'man' (e.g. *anthropos*, see 48), which addresses the different nuances in each of the terms and the languages involved, as well as methods to establish their original meaning and significance with a view to an accurate (*propria*) translation. Humphrey also sampled the opening paragraph of Aristotle's *Politics* in its Greek original, followed by his own translation and comments that illustrate the ways in which an original should be anatomised and interpreted before it can be rearranged into an accurate and efficient translation. It is revealing to find Humphrey discussing the translation of important concepts in Greek political philosophy against the background of his own political interests and contacts, for the translator here is faced with the challenge of finding Latin equivalents for Greek concepts such as *πόλις* (*civitas*), *κοινωνία* (*societas*), *ἀγαθός* (*bonum*), and the way they are combined in Aristotle's definition of civil society and public moral philosophy.[30] The vocabulary and definitions involved express basic principles

---

29 'Synonyma sunt, apud Graecos *ὄχλος, λαός, ὁι πωλλοί, δῆμος*, ut apud Latinos turba, multitudo seu *uulgus, & plebs: τοις πολλοις uulgus seu multitudinem dicerem, ὄχλου turbam, λαὸυ non plebem, sed populum.* Et merito uerterem interpretem suggillauit Laurentius Valla, quod plebem λαὸυ uerterit, quum populus uniuersos ordines Reipub. Contineat, plebs infima pars sit populi. Dein turba & rixa quantumuis idem significantia, ab Vlpiano disterminantur, alterum multitudinis esse tumultuantis, alterum duorum inter se contendentium & altercantium' (47–48; italics are mine).

30 The original Greek text is the following: 'ἐπειδὴ πᾶσαν πόλιν ὁρῶμεν κοινωνίαν τινὰ οὖσαν καὶ πᾶσαν κοινωνίαν ἀγαθοῦ τινος ἕνεκεν συνεστηκυῖαν (τοῦ γὰρ εἶναι δοκοῦντος ἀγαθοῦ χάριν πάντα πράττουσι πάντες), δῆλον ὡς πᾶσαι μὲν ἀγαθοῦ τινος στοχάζονται, μάλιστα δὲ [5] καὶ τοῦ κυριωτάτου πάντων ἡ πασῶν κυριωτάτη καὶ πάσας περιέχουσα τὰς ἄλλας. αὕτη δ' ἐστὶν ἡ καλουμένη πόλις καὶ ἡ κοινωνία ἡ πολιτική' (*Aristotle's Politics*, ed. W.D. Ross (Oxford: Clarendon Press, 1957), digital version available at http://www.perseus.tufts.edu/hopper/text?doc=Perseus%3Atext%3A1999.01.0057%3Abook%3D1%3Asection%3D1252a, accessed on 21 August 2022). And this is Humphrey's translation: 'Quoniam omnem ciuitatem esse societatem quondam uidemus, omnemque societatem boni alicuis causa constitutam: eius enim gratia quod bonum esse uidetur, omnia agunt omnes: perspicuum est cum omnes bonum aliquod habere propositum, tum uero summum omnium & principale eam quae omnium princeps est, & caeteras omnes complectit. Ea autem est quae ciuitas est nominate, ciuilisque societas.' (Humphrey, *Interpretatio*, 34). A recent English translation of the paragraph reads: 'Every state is a community of some kind, and every community is established with a view to some good; for everyone always acts in order to obtain that which they think good. But, if all communities aim at some good, the state or political community, which is the highest of all, and which embraces all the rest, aims at good in a greater degree than any other, and at the highest good.' (Aristotle's *Politics*, B. Jowett, trans., in *The Complete Works of Aristotle*, Vol. 2. The Revised Oxford Translation, ed. Jonathan Barnes (Princeton: Princeton University Press, 1984), 1986).

and establish nuanced distinctions between nouns that denote sophisticated concepts employed in the description and regulation of important aspects of a political community.[31] This example is of a piece with Humphrey's discussion of different ways of referring to the members of the political community in Greek, his interest in the morphological typology and the translation of vocabulary related to *bonum*, and the different ways of denoting humanity in Greek and Hebrew. This demonstrates, in short, Humphrey's awareness of the fundamental epistemological, and eventually normative, role of translation when it comes to establishing the vocabulary in which a specific discipline or a debate must be conducted and successfully argued. In this particular case it involves formulating and articulating the doctrines and principles that regulate private and public behaviour in the closely connected realms of religion, politics and moral philosophy.

The Translator as Proteus

In the opening pages of the *Interpretatio* ([a8^{r-v}]), Humphrey declares that the overall scope of his work includes teaching techniques for the acquisition of skills in foreign languages as the indispensable precondition for the practice of translation and interpretation. Indeed, the third and last book in Humphrey's *Interpretatio* consists of a series of practical drills in language learning through translation with an even more abundant profusion of examples (*De ratione interpretandi liber tertivs: progymnasmatum & exemplorum syluuam continens*). It sanctions Humphrey's text as a supreme example of the application of grammar and rhetoric to translation, not just as an abstract theory but as part of a very pragmatic educational project.

As is his wont at the onset of each major section, Humphrey's preface to this third book recapitulates the contents he has presented so far and proclaims the pedagogical intent of the work, which culminates in this section with its exercises in what he describes as *ars conuertendi* (413). These contents are then arranged in such a way that Humphrey proceeds in a systematic and orderly fashion, from the interpretation of the textual layout in original manuscripts to the lower levels in the codification of linguistic information – including examples and comments on morphological and syntactic equivalences between different languages – and then proceeding on to higher discurs-

31 'αὕτη δ' ἐστὶν ἡ καλουμένη πόλις καὶ ἡ κοινωνία ἡ πολιτική' is rendered by Humphrey as 'Ea autem est quae ciuitas est nominate, ciuilisque societas'.

ive levels, such as the use of figurative language, and also including alternatives to literal or close translation such as paraphrasis.[32]

The translator, Humphrey proclaims, is a Proteus. To this effect, these exercises and the principles that sustain them will provide not just basic linguistic skills but also the sort of expressive versatility which will enable him to speak like a prophet when rendering Scripture, or reproduce with as much fidelity as possible the voice of an orator, the style of a poet, or the narrative structure of a historian. The best translator must be capable of ventriloquising the Greek of Homer, Demosthenes, Thucydides or Menander in the translated text.[33] In order to do so, he must absorb the stylistic features (*ornamenta*), the figures of speech (*tropi*), even the rhythm and the cadence (*numeri*) of the original and reproduce them in the target language with due respect paid to *more* and *consuetudo*, i.e. the uses of the language into which he is rendering the original with as much truth (*lingua uere*) as he can muster.[34] But before achieving this supreme feat of *imitatio* of the language and the style, the *forma* and the *tenor* of the authors he is translating (to the point of standing in for them), the translator's approach to the text must start from a much lower and more pragmatic level of detail. Other than the rules of grammar and the precepts of rhetoric, this includes absolute familiarity with the script, the punctuation and in general any other paralinguistic components used in the manuscript in question (e.g. *characterem scriptionisque genus, abbreuiationes*).[35]

The third book becomes in large part a treatise in comparative morphology and syntax applied to the appropriate translation of all sorts of disciplines and genres, from theology and Scripture to literature and philosophy, with examples taken from Hebrew, Greek, Latin, English and other vernaculars. For example, Humphrey recounts how he gave a young English pupil one

32 See 442–445.

33 See 416–418.

34 'Simili cura *Rhetorica ornamenta, numeri, ac tropi* sunt cognoscendi, quibus oratio tota colorata est & ornate: quibus intellectis maxime iis quae sententiarum schemata dicunt, tota circumscriptio notanda & animo concipienda, ut tu *more* ac *lingua uere*, commode et quam poteris elegantissime, ac si iam tua esset, efferatur' (421–422).

35 'Alioqui si *characterem scriptionisque genus*, autorisque ingenium percusione hac breui & celeri exploratum non habueris, naufragium feceris, similiter ut nautae faciunt, qui percontando aunt contis tentando metiendoque exquirunt omnia, ante quam se mari temere periculoseque committant. Quod etiam in manuscriptis autographis faciundum est necessario, ut *abbreuiationes* illas quas uocat & nouas ac contractiores scribendi rationes praedicas, que in omnibus saepe, in Grecis maxime reperiuntur' (419–420, italics are mine).

of Cicero's letters to translate, and then he uses this translation to discuss techniques and common errors coming from an English-speaking beginner in Latin (as he had done earlier in the section dealing with German).[36] Then, for his 'exemplum Gallicum', Humphrey quotes a historical text to illustrate a case of interference between French and Latin when translating from the former to the latter. The 'Exemplum Hebraicum' concerns a couple of translations from Obadiah, whereas the 'Exemplum Graecum' uses a text by Philo of Alexandria (an author Humphrey knew well, as we have already seen).[37] The frequent phenomenon of interference between two languages involved in translation is a common problem of which Humphrey was well aware, and to which he had already devoted some attention in other sections.[38] On top of these practical examples that illustrate what we might call the nuts and bolts of translation proper, Humphrey also discusses other methods that can be used to interpret and render an original into another language in a way that maximises its understanding by the targeted readers. These include paraphrasis and metaphrasis, which he exemplifies with cases of famous classical and modern scholars who practised them, such as Cicero (530), Macrobius (531), Erasmus – who produced useful paraphrases of the *New Testament* – and Philip Melanchthon, who paraphrased Cicero himself (532).[39]

In defending the principle of what he calls *plena interpretatio* in his first book, Humphrey had resorted to terms and concepts and in general doctrines from Aristotle's *Rhetoric* and the Greek tradition in general to its Roman *translation* or reformulation in Cicero's Latin. Cicero's relevance is based not just on his intellectual prestige but also on the fact that he was the main conduit for the appropriation of classical Greek doctrines and the political capital they carried with them. Humphrey appears fully aware of his importance in terms of the doctrines he reformulated for readers of Latin, but also of his relevance as an example of how this sort of appropriation through translation, paraphrasis and rewriting could be actually performed. Cicero's texts were them-

---

36 'Exemplum Anglicum in Cic. Epistolis' (549–553). This English student might have been Edward Stafford (1552–1605), the son of another Marian exile, William Stafford, who died in Geneva in 1556. His widow and son did then move to Basel, where it seems Humphrey must have been this young boy's instructor. See the 'Exemplum Germanicum A. Schori' (547–549).

37 Further details at 554–560, 566–622.

38 See for example 541–542.

39 See 530–532, after which he moves on to the paraphrasis of poetry. Well beyond translation and paraphrasis, he also discusses here the concepts of *emulatio* as competition with the greats, which he illustrates with the work of Timolaus Larisseus (see 536–537), and also the names of Plato's Italian translators and emulators (536–537).

selves tapestries woven by the different ways in which he recast the wisdom of the Greeks to serve the purposes of universal Roman hegemony, one of whose harbingers he eventually became.

Translation, Knowledge and Power

Humphrey was perfectly aware of the notion of *translatio imperii*, and once more Cicero is explicitly presented as his model: a few paragraphs towards the end of his book (498 ff.) are devoted to a description of how Latin collected all the wisdom of the Greek world, with Cicero as one of the great agents in this process of appropriation. These are ideas that resonate with the principles of fifteenth-century humanism, its reassessment of grammar, rhetoric, the methods of textual archeology used in philology, and their joint application to translation as interpretation and imperial appropriation, which takes us back to Lorenzo Valla, another of Humphrey's favourite models. As he recounted in his dedicatory preface to his Latin translation of Thucydides, Valla was part of the team of translators who had set about rendering Greek classics for the Latin library that their patron, Tomasso Parentucelli, the humanist Pope Nicholas V, was putting together in Rome. Valla describes himself and his fellow translators as merchants who brought the wealth of Greek cultural products back home to the reborn Roman empire (a trope also used by Humphrey).[40] But Valla also described the Pope as a new emperor and his translators as the generals sent abroad to conquer Greek riches and take them back to Rome as the spoils of empire. There were other cases besides Valla that Humphrey knew very well, such as the member of the wealthy Fugger family who had sponsored the Hebrew dictionary in which he participated, and of course the Italian example of the Medici, their manuscripts and their famous library. In fact, before he was elected pontiff, Parentucelli had drawn up for Cosimo de' Medici (*Cosimo il Vecchio*, the founder of the Medici dynasty) a *canone* or catalogue of all the books necessary for a comprehensive library, and once he became pope he embarked on a project of translation of Greek classics into Latin, which was part of his ambitious plans for the expansion of the Vatican library. Humphrey also praises other European linguistic communities and nation states for the way they have enriched their languages and their countries with the knowledge they have translated into it. He finally devotes some space to how England has

40 In page 515 of the *Interpretatio* Humphrey compares the enrichment of the national language through translation with the wealth created by merchants who import exotic goods (*merces exoticas*) that are not originally available in the homeland.

also undergone a similar process, and to that effect he quotes a Neolatin poem by John Leland in praise of the Earl of Surrey and Sir Thomas Wyatt.

There is consequently an early imperialist dimension to Humphrey's treatise which could be described as the cultural politics of translation and language instruction. This appears most eloquently in Humphrey's defence and praise of philology and translation (*interpretatio*) as means to achieve universal knowledge and become familiar with the entire globe and all its riches, which he proclaims in his dedicatory preface.[41] That translation and interpretation are both related to the teaching and learning of languages are some of the ideas that frequently recur with variations throughout Humphrey's *Interpretatio*. They are all part of the toolbox of empire and therefore of a piece with other sorts of skills and virtual capital, in science and several other disciplines and practices, including transport, travel, cartography, architecture and engineering for the construction of infrastructures, and in general methods for the codification and communication of information and knowledge and for the conduct of trade.

Humphrey's networks once again cast some revealing light upon his *Interpretatio* as part of a wider joint effort by members of Queen Elizabeth's court to support the process of state-building at home that had been also taken up by other European nations. Like the other Marian exiles he met abroad, Humphrey was part of a committed group of future high-ranking officers, political, religious and ideological agents, who were getting ready during these years to take over as soon as they could return to England to re-establish Protestantism. As time went by, and above all after Mary Tudor's death in 1558, this prospect became increasingly possible, and one of the ways in which we

41 'Haec, ut breui complectar omnia, Mundum ostendit, Europae, Asiae, Aphricae diuitias explicando: ex Graecia liberalium disciplinarum circulum deportanto: noua, ritus, mores, instituta ex Germania, Gallia, Italia, Hispaniis ab extremis mundi partibus apportando' (a3^{r}). Baki Tezcan's chapter below on Ali Bey/Albertus Bobovius addresses the use of translation as a tool for Western imperialism, driven by Christian millenarism and Eurocentric universalism. It is revealing that these projects should include not just the translation of the Bible but also of entire multilingual dictionaries, which demonstrates the instrumental function of grammar and lexicography at the service of empire, with translators viewed as conquering generals sent abroad to bring back home the spoils of war. Both Valla and Humphrey constitute early examples of this world view and its attached practices and strategies, which in some cases – such as those described by Angelo Cattaneo in his chapter – facilitated the sort of cultural translation that was fundamental for Jesuit missionaries in their approach to the languages and belief systems of those they sought to convert. In contrast, Ann Thomson's essay analyses Joseph Morgan's efforts to counter the negative views of Islam fostered by those who had preceded him, although he was driven by a patriotically Protestant anti-Catholic agenda.

can understand Humphrey's *Interpretatio* is as part of this vast operation to take power and then legitimise it. This context explains his determination to lend his intellectual acumen to the mobilisation of knowledge at the service of the new regime, which includes translation as both knowledge and a method for appropriating political and doctrinal capital. Humphrey's project is not dissimilar, in this respect, to the project drafted in a manuscript by the explorer Humphry Gilbert: the so-called *Queene Elizabethes Achademy*. Composed around 1570, it proposed the creation of an institution of higher education for the teaching of foreign languages, the collection of information and the production of research and knowledge that could be directly or indirectly applied to facilitate England's imperial expansion through exploration, conquest and trade. The aim of it was to give the nation a competitive edge over other European powers who were at the time also engaged in their own projects of global colonisation.[42] Francis Walsingham himself, who as is well known was to become Elizabeth's spymaster, constitutes an egregious example of how information and intelligence could be weaponised and mobilised in the defence of England's interests at home and abroad, something which is of a piece with Humphrey's own anti-Catholic, Reformist propaganda. The *Interpretatio* also provides a very good case study for a vertical approach to translation and communication; i.e. one that does not just look into the origins of texts, their paths and destinations, but also the process of composition, from its most humble components (e.g. caligraphy, orthography, even the production of paper in different formats), the artisans involved in it, to the textual critics, editors, publishers and translators who constitute the complex team of human resources, the intellectual and material infrastructure that underlies the finished product.

Humphrey's years in Oporinus's and Froben's Basel exposed him to the culture of translation, publication, information and knowledge engineering that involved all the different techniques employed in the production of sophisticated texts; these texts then circulated throughout the rest of Europe in original critical editions, in Latin translation, and above all in the vernacular versions that gained new readers for the refreshed canon of classical texts and the Word of God in all the different European languages. As the preceding paragraphs have described, Humphrey was involved in many projects of this sort, and his

42 Humphrey Gilbert's manuscript was first printed in *Queene Elizabethes achademy / (by Sir Humphrey Gilbert). A booke of precedence. The ordering of a funerall, etc. Varying versions of the good wife. The wise man, etc.* edited by Frederick James Furnivall. With essays on early Italian and German books of courtesy / by W.M. Rossetti and E. Oswald (London: Early English Text Society, Extra Series 8, 1869) 7.

Interpretatio is an excellent example of the different components involved in the process and of the skills required and the methods employed. We could say that it was a handbook for the accurate appropriation of cultural, political and religious capital, as well as for the instruction of the new cadres who, in each of the emerging European nation-states, would be in charge of furnishing normative discourse. This would be done through the articulation of a body of texts and the invention of the sort of vocabulary which could sustain the complex hierarchies and ideological structures that intervened in the generation of national identities and the creation and legitimation of their new political and religious institutions.

As we have seen Humphrey presented himself as a new Ulysses, part of a fresh breed of religious, political and intellectual refugees who wandered now not across the Mediterranean as Homer's hero had done after the Trojan war but across European borders, driven into exile by the conflicts generated by the Protestant Reformation. Before him, and over the course of the late thirteenth and early sixteenth centuries, Greek-speaking refugees had escaped the gradual advances of the Ottomans in the Eastern Mediterranean and brought with them classical Greek manuscripts. Starting in places like Florence, they set about teaching the language and facilitating skills for their accurate interpretation, which laid the foundations not just for a revival of classical Greek authors but also for fresh approaches to the original texts of the New Testament. While Greek grammarians, philologists and translators had first crossed the Mediterranean from East to West, the fruits of their labour would subsequently intersect with another major wave of refugees coming from the opposite direction, crossing their paths in hubs of knowledge like Basel. The Greek that Lorenzo Valla had learned from Eastern Roman Hellenists in mid-fifteenth-century Italy facilitated his commentaries on the Greek Gospels that would be subsequently edited in print by Erasmus, who would also use them for his own Greek edition and Latin translation. Luther, in turn, would use Erasmus's Greek edition for his influential translation into German. The spirit and far-reaching consequences of this series of displacements of people, ideas and texts across the Mediterranean and then within European borders is encapsulated in the long subtitle to Erasmus's *Novum Instrumentum*, the second printed edition of the original Greek of the New Testament after the Complutensian Polyglot Bible of 1514. Frobenius published it in Basel in 1516, and after briefly explaining the method he had used to fix the text in print, Erasmus concluded by enjoining his audience to 'read, know and then judge' (*lege, cognosce, ac deinde judica*), a true harbinger of the critical spirit that would materialise again in Kant's famous *aude sapere*.

Acknowledgements

Research for this essay was made possible by visiting fellowships at Magdalen College, Oxford, in 2020 and 2023. I wish to express my gratitude to the college and its fellows for their generosity and hospitality.

CHAPTER 2

'Come Poeta Che Finge, Non Come Filosofo Che Argomenta'

Italian Translations of Fontenelle's Entretiens sur la pluralité des mondes (*ca. 1730–1751*)

Giovanni Lista

Introduction

From its first publication in 1686, Fontenelle's *Entretiens sur la pluralité des mondes* became one of the greatest bestsellers of the eighteenth century, undergoing countless translations and imitations across Europe. In a seductive dialogue reminiscent of Jean Racine's and Molière's elegant prose, Fontenelle's work displays a philosopher amusing a noblewoman with the astronomical demonstration of extra-terrestrial life across the solar system. The *Entretiens* ingeniously mocks Scholastic physics of Aristotelian origin to endorse the acceptance of the Copernican system and to defend the discoveries of Galileo and Kepler, while embracing Cartesian physics to explain the natural mechanisms ruling a secularised universe.

From the traditional perspective of the history of science, the *Entretiens*'s astonishing success depended less on it being a 'chef d'œuvre de vulgarisation' (masterpiece of popularisation) of scientific ideas than on Fontenelle's choice of a polite literary genre to convey the captivating image of a plurality of worlds to his European audience.[1] In 1687, Isaac Newton's *Philosophiæ Naturalis Principia Mathematica* and its law of universal gravitation discredited Fontenelle's version of Cartesian physics as a workable hypothesis, progressively dismissing most of the scientific contents of the *Entretiens*. While the social phenomenon of the 'astronomie fontenellisée' (Fontenellised astronomy) contributed to initiate a golden age of scientific dialogues in the eighteenth century,[2] the inherent tension between presenting the results of scientific observations and the open dialogical literary genre determined an epistemic and cognitive shift,

1 Christophe Martin, introduction to *Entretiens sur la pluralité des mondes* (Paris: Flammarion, 1998), 23–24.

2 Michel Foucault, *Les mots et les choses: une archéologie des sciences humaines* (Paris: Gallimard, 1966), 103.

 | DOI:10.1163/9789004731806_004

especially from the 1750s onwards. Newton's breakthrough promoted 'systems' as both the prevailing models of interpretation of reality and as the most common literary genre to disseminate scientific knowledge, relegating the dialogical form to its classical past.[3]

Since the nineteenth century, historiography has therefore reflected this vibrant cohabitation of a scientific discourse and a literary discourse in Fontenelle's *Entretiens*, until the most recent elaboration of quantum physics and multiverse theories brought back to the fore the work's emphasis on a plurality of worlds. In their eighteenth-century contexts however, one should resist considering the *Entretiens* as a proxy for the circulation of Cartesianism as a whole. To a certain extent, most of Fontenelle's work was 'un cartésianisme de polémique' (a polemical Cartesianism) employed to defend his idea of modernity,[4] a rhetorical tool to mark the watershed between ancient and modern philosophy within the history of science. In the *Entretiens* in particular, Fontenelle's interpretation detached Cartesian mechanism from its original system of philosophy and metaphysics. This led Fontenelle to reject the occasionalist readings of Descartes supported by the likes of Louis de La Forge and Nicholas Malebranche which, by attributing the original cause of motion to God's action, denied any principle of causality in physics. Rather, the *Entretiens* put forward a cosmology entirely based on Descartes's theory of vortices: for Fontenelle, it is the impacts and frictions of large spinning groups of material particles with each other that cause their motion, in a universe entirely made of interlocking matter.[5]

---

3 Charles Bazerman, *Shaping Written Knowledge: The Genre and Activity of the Experimental Article in Science* (Madison, WI: University of Wisconsin Press, 1988); Walter J. Ong, *Ramus, Method, and the Decay of Dialogue: From the Art of Discourse to the Art of Reason* (Chicago: University of Chicago Press, 2004); Fabrice Chassot, *Le dialogue scientifique au XVIIIe siècle: postérité de Fontenelle et vulgarisation des sciences* (Paris: Classiques Garnier, 2011); Clifford Siskin, *System: The Shaping of Modern Knowledge* (Cambridge, MA: MIT Press, 2017); Helge Jordheim, 'The Fontenellian Moment: Revisiting 16th and 17th Century Worlds', in Helge Jordheim and Erling Sandmo, eds, *Conceptualizing the World* (New York: Berghahn Books, 2018).

4 Simone Mazauric, *Fontenelle et l'invention de l'histoire des sciences à l'aube des Lumières* (Paris: Fayard, 2007), 229.

5 Marie-Françoise Mortureux, 'Les "Entretiens sur la pluralité des mondes" de Fontenelle: Discours scientifique, discours littéraire', *Littérature*, no. 4 (1971): 44–52; Steven Nadler, *Causation in Early Modern Philosophy: Cartesianism, Occasionalism, and Preestablished Harmony* (University Park, PA: Penn State Press, 1993); Lisa Downing, 'Occasionalism and Strict Mechanism: Malebranche, Berkeley, Fontenelle', in *Early Modern Philosophy: Mind, Matter, and Metaphysics*, ed. Christia Mercier and Eileen O'Neill (Oxford: Oxford University Press, 2005), 206–230; Tad M. Schmaltz, 'Occasionalism and Mechanism: Fontenelle's Objections to Malebranche', *British Journal for the History of Philosophy* 16, no. 2 (May 2008): 293–313; Mitia Rioux-Beaulne, 'What Is Cartesianism? Fontenelle and the Subsequent Construction of Cartesian

To address the issue of what forms of knowledge circulated in Italy by way of translations of the *Entretiens*, the present study will make use of two complementary methodologies which elaborate on the idea, central to this volume, that the mobility of both texts and their translators should constitute an essential element of translation studies. The first methodological approach seeks to redefine the medium of dissemination, that is the text itself. In looking at some translations of the *Entretiens*, Helge Jordheim has recently replaced Bruno Latour's notion of printed media as vehicles of circulation of knowledge with the dynamic concept of texts as sites of knowledge. In its flexible dialogical form, Fontenelle's work constituted an open and prolific site of circulation of different kinds of knowledge; depending on the cultural and temporal contexts of its different English, Danish or German editions, the *Entretiens* transferred information in several fields, ranging from anthropology and linguistics to theology and astronomy. From this angle, the Italian translations studied here reveal additional information that complicates any teleological narrative related to the history of science.[6]

The second approach adopted in this chapter considers the actors intervening in this site of knowledge and defining through their own agency and agendas which knowledge is primarily transmitted; in other words, the translators themselves. As the research of Peter Burke, Laszlo Kontler or Fania Oz-Salzberger has shown, while the circulation of translations can be taken as a quantitative indicator of the success of a publication, translation practices have a significant impact on the quality of the reception of any textual production. Accordingly, the intentions, actions and biographical backgrounds of the translators become decisive in shaping the contents of the translation, as the outcome of a negotiation between the 'source' culture and the 'target' culture. It is therefore necessary to look at translators as cultural mediators and even ideological gatekeepers, who sometimes implant additional knowledge in the original text through all sorts of paratextual devices, such as prefaces, footnotes, commentaries or afterwords. This often resulted in substantial displacements of meanings, as is the case with the Italian translations at the centre of this study.[7]

Philosophy', in *The Oxford Handbook of Descartes and Cartesianism*, ed. Steven Nadler, Tad M. Schmaltz and Delphine Antoine-Mahut (Oxford: Oxford University Press, 2019), 481–496.

6 See Bruno Latour, 'Visualisation and Cognition: Drawing Things Together', *Avant: Trends in Interdisciplinary Studies*, n°3 (2012): 207–260; Helge Jordheim, 'The Printed Work as a Site of Knowledge Circulation. Dialogues, Systems, and the Question of Genre', in *Circulation of Knowledge: Explorations in the History of Knowledge*, ed. Johan Östling et al. (Lund: Nordic Academic Press, 2018), 541–593.

7 See Fania Oz-Salzberger, 'The Enlightenment in Translation: Regional and European Aspects',

On these premises, this essay advances the first comprehensive study of the Italian translations of the *Entretiens* in the first half of the eighteenth century, focussing on the role of the translators as much as on the textual translations they produced.[8] While the controversy between Cartesian science and Newtonian ideas naturally constitutes one of the frameworks for this investigation, I will mainly consider the written production and intellectual activities of the translators as the main tools to determine what kind of knowledge they sought to transmit with their versions of the *Entretiens*. After a brief survey of some early manuscript translations, pivotal to my undertaking will therefore be rank and file figures of the Italian *Settecento* (eighteenth century). I will first look at the Neapolitan abbot and lexicographer Annibale Antonini (1702–1755), who edited the *Ragionamenti sù la pluralità de' mondi* in 1748. Antonini taught Italian to Parisian noblewomen and was an active promoter of Italian literature: in his preface, he gave the credit for the literary translation to one of his brilliant pupils. Secondly, I will move to the cleric Bernardino Vestrini (1715–1786), who as a member of the Etruscan Academy in Cortona published his own version of the *Entretiens* in 1751. Interested in physics and astronomy, Vestrini employed a strictly scientific approach to Fontenelle's text, engaging with its Cartesian cosmology in his afterword.

Early Manuscript Translations and Their Contexts (ca. 1698–1730)

While the circulation of Fontenelle's *Entretiens* and their translations in Italy was certainly affected by early censorship, they were well known across the Peninsula. In 1687, the academician in Rome and consultant for the Congregation of the Index Giovanni Giustino Ciampini wrote the report adding them to the Catholic Church's list of prohibited books. Ciampini's motivations dogmatically mention heliocentrism and Copernicanism, but mostly insist on the

---

European Review of History: Revue Européenne d'histoire 13, n°3 (September 2006): 385–409; Peter Burke and R. Po-chia Hsia, eds., *Cultural Translation in Early Modern Europe* (Cambridge, MA: Cambridge University Press, 2007); László Kontler, 'Translation and Comparison, Translation as Comparison: Aspects of Reception in the History of Ideas', *East Central Europe* 36, n°2 (January 2009): 171–199; Stefanie Stockhorst, ed., *Cultural Transfer through Translation: The Circulation of Enlightened Thought in Europe by Means of Translation* (Leiden: Brill, 2010); Patrice Bret and Jeanne Peiffer, eds, *La traduction comme dispositif de communication dans l'Europe moderne*, (Paris: Hermann, 2020).

8 A partial discussion of the Italian case study can be found in the *Journal of the History of Ideas Blog*, https://www.jhiblog.org/2024/07/15/on-translating-knowledge-in-the-enlightenment-the-case-of-fontenelles-plurality-of-worlds/

theological contradiction of humans inhabiting other planets and obtaining redemption from Adam's original sin. While the hypothesis of a plurality of worlds was the heresy which kept the *Entretiens* on the Index until 1948, their Cartesian vortical system was still debated in Rome in the 1720s. In his *Pancosmosophia*, on which he was working in 1710, the Jewish philosopher in Venice Sabato Ambron rather chose astronomical grounds to refute modern scholars 'che la pluralità de' mondi suppongono' (who presume the plurality of worlds) and to advance his own model of the universe.[9] At the University of Padua, the mathematician Giovanni Poleni tried to mediate between the systems of Descartes and Newton in his *De vorticibus coelestis Dialogus* published in 1712. He sent a copy to his colleague in Berne Johann Bernoulli, who showed his appreciation by comparing Poleni's entertaining work to Fontenelle's.[10]

The existence of two manuscript translations confirms this early interest in the *Entretiens*. According to Rosanna Immacolato, a first and largely incomplete version can be dated around 1695 and consists of some 130 folios including Fontenelle's *Dialogue des morts* (1683).[11] The manuscript was preserved at the Jesuit College of the city of Catanzaro, in the former Kingdom of Naples, now hosting the high school library where it remained until the 1970s before going missing. Based on the lexical choices, Immacolato speculates that the author, Costantino Piscopo, was a Jesuit from Southern Italy with some previous experience as a translator and that besides being quite literal, Piscopo's version mediates between the original text and the need to capture the attention of an Italian reader. In one instance Piscopo replaces Fontenelle's original calculation of the distance the Sun and other stars would have to cover if they were to orbit around Earth with the beguiling image of these astral bodies being 'più lontane che dà quì alla Cina' (further away than from here to China); however, this change is taken from the 1698 French reissue, where Fontenelle

9 Ambron's treatise is now lost, see *Giornale de' Letterati d'Italia*, vol. 2 (Venice, 1710), 521–523.

10 Universitätsbibliothek Basel, L Ia 675:Bl.17, Johann Bernoulli to Giovanni Poleni, 8th July 1712; Giovanni Poleni, *De vorticibus coelestibus dialogus* (Padua, 1712); Margherita Palumbo, '"Ma qui si tratta anche del mondo nella luna …": La censura de "Le monde dans la lune" di John Wilkins', *Bruniana & Campanelliana* 17, no. 2 (2011): 501–511; Maria Pia Donato, '"Il patente disprezzo a cui non possiamo ormai fare più fronte": Benedetto XIV, le scienze e l'Inquisizione tra necessità apologetica e desiderio di "modernazione"', in *Storia, medicina e diritto nei trattati di Prospero Lambertini Benedetto XIV*, ed. Maria Teresa Fattori (Rome: Edizioni di Storia e Letteratura, 2013), 125–138; Gabriel Maugain, 'Fontenelle et l'Italie', *Revue de Littérature Comparée* 3 (January 1923): 541–603.

11 Biblioteca del Liceo Ginnasio P. Galluppi, Catanzaro, MS. 76-2-3, *Dialogi de' Morti del Sig. de Fontenelle, e, i Trattenimenti su la pluralità de' Mondi del medesimo tradotti dal francese da Costantino Piscopo*, ca. 1695.

himself employed this comparison instead of reprinting the erroneous estimates contained in the 1686 edition. The manuscript ends abruptly after the second evening suggesting the moon may be inhabited, and might have been intended only as an exercise by the translator, possibly as part of the College's curriculum.[12]

Even if composed after 1698, Piscopo's *Trattenimenti* are interesting evidence of an Italian translation underway before other European versions – such as the Danish or the Dutch – were published. Although the loss of the manuscript makes it difficult to fully reconstruct its context, the Kingdom of Naples had been a fertile laboratory for a materialist reception of Cartesian philosophy since the mid seventeenth century. In a vein similar to Fontenelle's, Neapolitan scholars such as Pietro Giannone and Francesco D'Andrea developed a scientific methodology that dispensed with Descartes's metaphysics and focused on mechanism. This allowed them to question Scholastic physics without entering philosophical debates that would have invited the attention of the Inquisition, and to praise Descartes as the harbinger of a new science. The Cartesian theory of vortices central to the *Entretiens* was a matter of contention until Newtonian science came to dominate Naples *accademie* (academies) from the 1730s.[13] A case in point is Celestino Galiani's *Epistola de gravitate et cartesianis vorticibus* (1714), which replied to the *renatista* (Cartesian) Gregorio Caloprese by refuting Descartes's vortical system.[14] In the second half of the eighteenth century, Fontenelle would gain popularity in Southern Italy, and new translations of the *Entretiens* would appear as part of his collected works in 1765, and again separately in 1784 and 1788.[15]

---

12 Compare Bernard Le Bovier de Fontenelle, *Entretiens sur la pluralité des mondes. Nouvelle édition, augmentée d'un nouvel entretien* (Paris, 1698), 46 with the extract in Rosanna Immacolato, 'Un'antica traduzione da Fontenelle conservata in un fondo manoscritto calabrese', in *Civiltà di Calabria: Studi in memoria di Filippo De Nobili*, ed. Augusto Placanica (Chiaravalle: Effe Emme, 1976), 151–154.

13 See for instance Costantino Grimaldi, *Risposta alla terza lettera apologetica contra il Cartesio creduto da più d'Aristotele, di Benedetto Aletino* (Naples, 1703) and the translation of Descartes's *Principia Philosophiae* (1644) by Giuseppa Eleonora Barbapiccola, *I Principi della filosofia di Renato Des-Cartes* (Naples, 1722).

14 In the 1770s, the Italian poet Pietro Metastasio would reminisce how during his education as a young man Caloprese 'mi conducea quasi per mano fra i vortici dell'allora regnante ingenoso Renato' ('led me as if by the hand through the vortices of the then-reigning ingenious René'), see Pietro Metastasio, *Lettere*, vol. I (Nice, 1786), 127.

15 Celestino *Galiani, Epistola de gravitate et cartesianis vorticibus* (Naples, 1714), in Vincenzo Ferrone, 'Celestino Galiani e la diffusione del Newtonianesimo. Appunti e documenti per una storia della cultura scientifica del primo Settecento', *Giornale critico della Filosofia Italiana* 2, no. 1 (1982): 19–33; Pierre Girard, 'The Materialist Reception of the

A second anonymous manuscript from 1730, stored in the archives of the University of Bologna, may be directly connected to local scholarly debates.[16] Praised by Fontenelle himself who later became an honorary member, the Academy of the Institute for Sciences in Bologna constituted the most advanced centre of scientific research in the Peninsula and was internationally renowned for the quality of its experiments. Its secretary Francesco Maria Zanotti had been the first to introduce 'la spiegazione de' vortici di Des Cartes' ('the explanation of Descartes's vortices') in his lectures and became the teacher of the polymath Francesco Algarotti,[17] whom Zanotti introduced to the *Entretiens* in 1728 together with the astronomer Eustachio Manfredi. Algarotti's *Newtonianesimo per le dame* (1737) later openly emulated Fontenelle's dialogical model and gave the Venetian polymath European recognition at a time when both Zanotti and Manfredi had converted to Newtonian physics and optics in particular. However, as in Naples, Descartes's celestial *tourbillons* (vortices) were a subject of debate both in private correspondences and at the Bologna Academy until the 1730s.[18] The passionate supporter of Cartesian philosophy Laura Bentivoglio Davia, for instance, still defended the vortex model against the rising star of Laura Bassi, a graduate of the University who later held the chair of physics and completed the shift to a curriculum based on Newtonian theories.[19]

Cartesian Physics in Naples', in *Physics and Metaphysics in Descartes and in His Reception*, ed. Delphine Antoine-Mahut and Sophie Roux (New York: Routledge, 2018); Giulia Belgioioso, 'Images of Descartes in Italy', in *Receptions of Descartes: Cartesianism and Anti-Cartesianism in Early Modern Europe*, ed. Tad M. Schmaltz (Routledge, 2005); Claudia Addabbo, 'The *Philosophiae Naturalis Principia Mathematica* in Naples', in *Reading Newton in Early Modern Europe*, ed. Elizabethanne A. Boran and Mordechai Feingold (Leiden: Brill, 2017), 21–63.

16 Biblioteca Universitaria, Bologna, MS. 3768, *Trattenimenti sopra la Pluralità de Mondi Del Signor Fontanelle … Trasportati dall'Idioma Francese nell'Italiano da R.G. nel 1730*.

17 Francesco Maria Zanotti, *Opere di Francesco Maria Cavazzoni Zanotti* (Bologna, 1779), xxv.

18 See the commentaries of the Academy, where a study from the astronomer Giuseppe Antonio Nadi trying to combine Cartesian vortices with the Newtonian system was presented and discussed, *De Bononiensi Scientiarum et Artium Instituto atque Academia Commentarii* (Bologna, 1731), 271 and *passim*.

19 Francesco Algarotti, *Il Newtonianismo per le dame, ovvero dialoghi sopra la luce e i colori* (Naples, i. e. Venice 1737); See also the Canto III of the *Paradiso* by Eustachio Manfredi, in *Rime del dottore Eustachio Manfredi* (Bologna, 1713), 50–54, which describes the sky and its planets using the vortex metaphor indebted to Descartes. For context, see Walter Tega, '*Mens agitat molem*, L'Accademia delle scienze di Bologna (1711–1804)', in *Scienza e letteratura nella cultura italiana del Settecento*, ed. Renzo Cremante and Walter Tega (Bologna: Il Mulino, 1984); Marta Cavazza, *Settecento inquieto: alle origini dell'Istituto delle scienze di Bologna* (Bologna: Il Mulino, 1990), 240–243; Paula Findlen, 'Women on the Verge of Sci-

In the elegant Bologna manuscript, the translator took the risk of reproducing the integrity of the original work and of emphasising its scientific contents, suggesting the intention of preserving and circulating them. The opening page displays a coloured astral map drawn after Juan Olivar's engraving in the 1686 French edition, which illustrates the universe as a plurality of worlds held together by a system of Cartesian vortices. In replicating this heliocentric Copernican visual paradigm, the illustrator enhanced the original figurative motifs by magnifying the anthropomorphic faces appearing on each star and planet (see fig. 2.1 below). In addition, the text is adapted from the 1714 The Hague edition and retains some controversial passages that would disappear from other published Italian versions: in the second evening, Fontenelle translated Ludovico Ariosto's *Orlando Furioso* (1516), to describe how during his journey to the moon Astolfo found all the things that had been lost on earth, including the alms distributed after death – which implied the futility of intercession for the departed. Shielding himself behind Ariosto again, two pages later Fontenelle ridiculed the Donation of Constantine, a forged document that had constituted the basis of the Pope's temporal power.[20]

Unfortunately, the provenance of this manuscript and how it arrived in the Papal States still remain unknown, for the University only acquired it at the death of its original owner Ubaldo Zanetti in 1769. A pharmacist by trade, Zanetti was an erudite bibliophile who possessed a private library containing more than 3,000 volumes, including scientific books, anticlerical pamphlets and around 700 manuscripts. Such a collection constituted an attraction for Bologna's local scholars, but was also open to an extended network of learned visitors from all over Italy like Algarotti himself. Zanetti's activities likewise led him to buy, sell or exchange volumes, mostly in Northern Italian cities such as Venice or Ferrara. In 1748, Zanetti was arrested for the possession of forbidden books and obscene political satire, and most of his private library was confiscated by the municipal authorities. A year later, from the Papal prison of Forte Urbano Zanetti asked his brother Giuseppe to send him the manuscript

ence: Aristocratic Women and Knowledge in Early Eighteenth-Century Italy', in *Women, Gender and Enlightenment*, ed. Sarah Knott and Barbara Taylor (Basingstoke: Palgrave Macmillan, 2005), 265–287.

20 Lucía Ayala, 'Cosmology after Copernicus: Decentralisation of the Sun and the Plurality of Worlds in French Engravings', in *The Making of Copernicus*, eds. Wolfgang Neuber, Thomas Rahn, and Claus Zittel (Leiden: Brill, 2015), 201–226; Bernard Le Bovier de Fontenelle, *Entretiens sur la pluralité des mondes*, ed. Christophe Martin (Paris: Flammarion, 1998), 90–91; Enea Balmas, 'Fontenelle en Italie', in *Fontenelle: Actes Du Colloque Tenu a Rouen Du 6 Au 10 Octobre 1987*, ed. Alain Niderst (Paris: Presses Universitaires de France, 1989), 577–588.

FIGURE 2.1 MS. 3768, c. [II]r, courtesy of Alma Mater Studiorum Università di Bologna. Biblioteca Universitaria di Bologna

translation of the *Entretiens*, perhaps to show it to the local castellan or to the noblemen visiting him. While it remains possible that the manuscript had been prepared by one of the members of the Academy in Bologna and given to Zanetti as a gift, the latter could have acquired it on one of his trips to Venice and brought it back to Bologna at the request of a scholar.[21]

Annibale Antonini's *Ragionamenti* (1748) as a Piece of Italian Literature?

It was only in 1748 that the first Italian translation eventually appeared. Its title page indicates 'Parigi' (Paris) as the place of publication and ostensibly claims to be the product of Bernard Brunet's printing press. At the time, Brunet was reissuing the last volumes of Fontenelle's complete works, and would a few months later become the official printer of the Académie française. In fact, Brunet's 1742 edition of the *Entretiens*, which includes the fable of the bees at the end of the third evening and is considered to this day the classic edition of reference, constitutes the original text for the translation. In October, the *Journal des sçavans* mentioned the publication of the *Ragionamenti sù la pluralità de' mondi*, specifying that 'Mr. l'Abbé Antonini a pris soin de l'édition' ('abbot Antonini supervised the edition').[22] In his preface, Antonini praises Fontenelle's work, 'nell'Italiano idioma tradotti da altra valorosa Dama Francese' ('translated into the Italian language by another talented French Lady'),[23] but does not reveal the name of the translator. The *Ragionamenti* do not appear to have circulated extensively either in Paris or in Italy, although the royal censor and prominent collector of Italian literature Albert-François Floncel possessed one of the relatively few existing copies in his library.[24]

By the time Antonini edited the *Ragionamenti*, he had become an internationally renowned grammarian and philologist, his Italian textbooks being

21 Biblioteca Universitaria, Bologna, MS. 3924, 7, XVII, Giuseppe Zanetti to Ubaldo Zanetti, 25 February 1749. For context, see Rita De Tata, *All'insegna della fenice: vita di Ubaldo Zanetti speziale e antiquario bolognese (1698–1769)* (Bologna: Comune di Bologna, 2007).

22 *Journal des sçavans* (Paris, 1748), 639.

23 Bernard Le Bovier de Fontenelle, *Ragionamenti sù la pluralità de' mondi* ('Paris', 1748), II.

24 Bernard Le Bovier de Fontenelle, *Œuvres de Monsieur de Fontenelle*, 8 vols. (Paris, 1742–1751); Jean-Gabriel Cressonnier, *Catalogo della libreria Floncel* (Paris, 1774), 67; Paul Delalain, *Les Librairies & imprimeurs de l'Académie française de 1643–1793: notices biographiques* (Paris, 1907), 89–94; Frédéric Barbier, Sabine Juratic, and Annick Mellerio, *Dictionnaire des imprimeurs, libraires et gens du livre à Paris: A–C* (Geneva: Librairie Droz, 2007), 331–332.

celebrated by Carlo Goldoni as the quintessential reference for Parisian courtiers.[25] Son of the Baron of San Biagio Alfonso, Annibale Antonini was born into a wealthy family of landowners in 1702, near Salerno. He first studied humanities in Naples with his older brother Giuseppe, who would later inherit his father's peerage, and then moved to Rome, where he completed his education and took his vows. In 1721, Antonini edited a volume of collected poems for the wedding of Pope Innocent XIII's nephew Marc'Antonio Conti and the noblewoman Faustina Mattei, personally collecting compositions from the likes of Paolo Rolli, Pietro Metastasio and Giambattista Vico.[26] After this publication, we know nothing about Antonini's activities and movements: some biographers suggest that he left for an extensive tour leading him to Germany, the Netherlands and England, although Giuseppe believed Antonini had left directly for France as soon as 1722. Later on, Giuseppe would complain he had lost trace of his brother and send a detailed physical description to the numismatist Matteo Egizio, asking him to search for Antonini in Paris.[27]

By 1726, Antonini was indeed living in the parish of Saint-Sulpice, and had started his career as *maître d'italien* (Italian teacher) for the offspring of the international upper middle class and aristocracy of the Paris salons. His rise to fame was the result of the simultaneous publication of a series of popular Italian grammars – *Traité de la grammaire italienne* (1726), *Grammaire italienne à l'usage des Dames* (1728), *Grammaire italienne pratique et raisonnée* (1746) – and bilingual dictionaries. As a devoted supporter and propagator of Italian culture, Antonini also revised and reissued several classics in Paris, ranging from

25 Goldoni cited Antonini in a stanza of the sonnet *Piccola Venezia*: 'El linguaggio italian, con mio contento / Caro diventa alla nazion francese, / E tutti i cortigiani e i parigini / Cerca maestri e compra l'Antonini' ('The Italian language, to my delight / Becomes dear to the French nation / And all the courtiers and Parisians / Look for teachers and buy the Antonini'). See Carlo Goldoni, *Delli componimenti diversi di Carlo Goldoni* (Venice, 1764), 240, quoted in Gianfranco Folena, *L'italiano in Europa: esperienze linguistiche del Settecento* (Einaudi, 1983), 406–407.

26 *Rime per le nozze dell'eccellentissimo signore D. Marc'Antonio Conti duca di Guadagnolo, e nipote della Santita di Nostro Signore Papa Innocenzo decimoterzo felicemente regnante. Coll'eccellentissima signora D. Faustina Mattei de duchi di Paganica dedicate alla medesima da Annibale Antonini Napoletano* ([Rome], 1722). Antonini also asked Ludovico Muratori for a contribution, but could not convince him: Biblioteca Estense Universitaria, Modena, Fondo Muratori, A.M.49.63, Annibale Antonini to Ludovico Muratori, 27th September 1721.

27 Biblioteca Nazionale Vittorio Emanuele III, Naples, Carte Matteo Egizio, MS XIIIC90/53, 1^{r-v}. and 90/54, 1^{r}.–v., Giuseppe Antonini a Matteo Egizio, 22nd February and 27th March 1736 and MS XIII.C.90/49, 1^{r}.–v., 'Memoria per lo Sig. Matteo Egizzio'; Eustachio D'Afflitto, *Memorie degli scrittori del Regno di Napoli*, vol. I (Naples, 1782), 389–390; Anna Buiatti, 'Antonini, Annibale', in *Dizionario Biografico degli Italiani*, vol. 3, (Rome, 1961), 517–518.

Giovanni Della Casa and Torquato Tasso to Ludovico Ariosto.[28] In his elaborate prefaces, Antonini dedicated his works to influential figures from all around Europe and built a network that also encompassed numerous Italian *letterati* (men of letters). The Neapolitan philosopher Paolo Mattia Doria solicited Antonini to intercede with the Académie française, hoping that they would publish his treatises in Paris, while the editor of the Lucca edition of the *Encyclopédie* Filippo Venuti gifted him with a copy of his Italian edition of Louis Racine's *La Religion* (1748). Under the patronage of Michele Giuseppe Morei (1743–1766), Antonini eventually became a member of the prestigious literary Accademia dell'Arcadia as 'Sennuccio Lucanico'.[29]

In her study of Antonini's grammars, Helena Sanson noted the evolution from the strict conservative literary approach displayed in the 1726 *Traité* to the simplified version contained in the 1728 manual *à l'usage des Dames* (for the use of Ladies). Much to some of his acquaintances' disappointment and mistrust,[30] Antonini turned from reasserting the primacy of the rigid Tuscan tradition with examples drawn from Dante, Petrarch and Boccaccio to a more essential set of rules for a living Italian language. In order to reach a wider public with no prior knowledge of Latin or Italian, Antonini initiated the ste-

28 The complete list includes Giovanni Della Casa, *Prose e rime di Messere Giovanni della Casa, edizione nuova. Riveduta, e corretta per l'Abbate Annibale Antonini* (Paris, 1727); Giovanni Battista Guarini, *Il Pastor fido, tragicommedia pastorale del cavalier Guarini, edizione nuova, riveduta e corretta per l'abbate Antonini* (Paris, 1729); Torquato Tasso, *Aminta favola boscareccia di Torquato Tasso. Edizione nuova. Riveduta, e corretta per l'abbate Antonini* (Paris, 1729); Torquato Tasso, *La Gierusalemme liberata. Poema eroico di Torquato Tasso* (Paris, 1744); Giovanni Giorgio Trissino, *L'Italia liberata da' Goti; riveduta, e corretta per l'Abbate Antonini* (Paris, 1729); Ludovico Ariosto, *Orlando furioso, di Ludovico Ariosto (pubblicato dall' abate Anuibale Antonini e preceduto dalla vita di L. Ariosto)*, 3 vols. (Paris, 1746).

29 For a general study on Antonini's production, see Marina Marietti, 'Un maître d'italien à Paris au XVIIIe siècle: l'abbé Antonini', *Parola del Testo* XIX, no. 1/2 (2015): 69–82. Paolo Mattia Doria to Annibale Antonini, 1742?, in *Manoscritti napoletani di Paolo Mattia Doria*, vol. 2, ed. Marilena Marangio (Galatina, 1979), 157–183. Biblioteca Città di Arezzo, *Miscellanea erudita ed amena dell'Abate F. Venuti*, MS. 446/11, fol. 251. Biblioteca Angelica, Rome, Archivio dell'Arcadia, Catalogo 6, c. 9^{r}. and Catalogo 5, c. 259^{v}. Additional correspondence with Italian men of letters like Giovanni Battista Fagiuoli and Anton Francesco Gori can be found in Florentine archives. See Biblioteca Riccardiana, Riccardiano, 3439/5, 297^{r}.–v. and Biblioteca Marucelliana, hereafter BMF, MS. B VII 1, fol. 506^{r}.

30 According to the Tuscan teacher in Paris Giuseppe Buonamici, Antonini had admitted his Italian and his grammars to be of poor quality, to the point that Buonamici would describe him as a complete fraud, see Österreichische Nationalbibliothek, Vienna, Sammlung Bartolommeo Gamba, Autogr. 43/70–71, Giuseppe Buonamici to Antonio Conti, 1727. See also Apostolo Zeno to Caterino Zeno, 26th June 1728, in *Lettere di Apostolo Zeno cittadino veneziano istorico e poeta cesareo*, vol. III (Venice, 1752), 517–518.

reotypical genre of grammars 'for ladies' – a category he associated with that of 'illiterates' or 'absolute beginners' – , which would enjoy a wide diffusion in the second half of the eighteenth century. In the process, his grammatical explanations shifted from promoting an erudite understanding of the Italian language and literature to privileging the sociability of the salons that they were now designed to facilitate. A case in point are the thirteen dialogues added to the 1728 handbook, to furnish the readers with polite formulas to be used regardless of their actual ability to speak the language. Telling his audience that learning Italian should be undertaken 'pour s'amuser' (to enjoy oneself),[31] Antonini acknowledged that his early critics were right in asking 'Qui a jamais vû [...] une Grammaire sans Entretiens?' ('Who has ever seen a Grammar without Dialogues?').[32]

If we turn to the *Ragionamenti* with this context in mind, it appears that Antonini's and Fontenelle's *Entretiens* share the display of a gender specific commitment to the transfer of knowledge.[33] Whereas Antonini sought to codify Italian as a language for courteous conversations, Fontenelle employed the fiction of a gallant dialogue to display the 'politesse' of Cartesian astronomy. Both insisted on conveying information in an entertaining form, accessible to uneducated ladies, to spare them the effort of engaging with pedantic erudition.[34] To accommodate Fontenelle's original intention, the translation is fairly elegant and overall faithful to that of the *Entretiens* in most of its controversial parts: Descartes's mechanism and vortices, Copernican heliocentrism and the inhabitants of the moon all figure in their Italian guise.[35] This rendi-

31 Annibale Antonini, *Grammaire italienne à l'usage des dames avec des dialogues et un traité de la poésie, par M. l'abbé Antonini* (Paris, 1728), I.

32 Antonini, *Grammaire*, 241. Helena Sanson, '"Simplicité, clarté et précision": Grammars of Italian "pour les Dames" and other Learners in eighteenth- and early nineteenth-century France', *The Modern Language Review* 109, no. 3 (2014): 593–616; Helena Sanson, 'Women and Language Codification in Italy: Marginalized Voices, Forgotten Contributions', in *Women in the History of Linguistics*, ed. Wendy Ayres-Bennett and Helena Sanson (Oxford: Oxford University Press, 2021), 59–90; Gabriele Beck-Busse, 'À propos d'une histoire des "Grammaires des Dames". Réflexions théoriques et approches empiriques', *Documents pour l'histoire du français langue étrangère ou seconde*, no. 47–48 (June 2012): 1–22; Norma Romanelli, 'Le grammatiche di italiano per francesi di Annibale Antonini (1726–1746)', *Studi italiani di linguistica teorica e applicata* XLVI, no. 2 (2017): 303–320.

33 It has been convincingly argued that Antonini's engagement with a female readership was part of a marketing strategy, but his belief that women should receive an education is repeatedly asserted elsewhere. See the Preface to Tasso, *La Gierusalemme liberata* and Andrew Ramsay, *I Viaggi di Ciro tradotti dall'idioma francese* (Naples, 1753?), esp. XIX–XXII.

34 On this issue, compare Fontenelle, *Entretiens*, 70, 135 with Antonini, *Grammaire*, II.

35 See *Ragionamenti*, IX: 'tradotta voce per voce, e col medesimo giro dell'idioma, in cui le

tion however implied a particularly challenging adaptation, as in the preface Antonini remarks that when compared to Italian, the French language is better suited to discuss scientific topics, since it is 'copiosa di Termini scientifici' (rich in scientific Terms).[36] A trace of the translation process can be found in Antonini's dictionaries, where the definition of the word *tourbillon* (vortex) changes from one edition to the other. While in the 1735 edition the entry is simply defined through the example of a whirlwind or waterspout, in 1743 Antonini decided to incorporate the Cartesian meaning of the word in his trilingual glossary:

> en termes de Philosophie, se dit D'un amas de matière, dont les parties se meuvent dans le même sens, & autour du même centre. (Lat. vortex.) *Vórtice*. Le système des tourbillons; le tourbillon du Soleil
>
> (in Philosophical terms, is said of a mass of matter, whose parts move in the same direction, & around the same centre. The vortex system; the vortex of the Sun).[37]

In fact, rather than a scientific work, the *Ragionamenti* are presented as a translation exercise brilliantly completed by one of Antonini's pupils; this further substantiates the primary agenda of demonstrating the superiority of 'la Lingua Italiana più d'ogni altra propria per tradurre' ('the Italian Language, more than any other appropriate for translating'),[38] because of the flexibility and breadth of its vocabulary. The astral map reproduced in the Bologna manuscript, for instance, is replaced in this edition by the allegory of a *femme savante* (learned woman) allegedly in the act of translating the *Entretiens* (fig. 2.2). In the same

cose furono scritte da prima' ('translated word for word, and with the same turn of phrase of the language in which they were originally written'). Some minor changes however suggest a certain degree of censorship: Antonini judiciously removed the reference to the Donation of Constantine, 74–75. A copy held at the Bayerische Stadtbibliothek München had the whole section referring to the alms for the dead removed, while another copy with the integral text is preserved at the Taylor Institution Library in Oxford.

36 *Ragionamenti*, VI.

37 Annibale Antonini, *Dictionnaire françois, latin, et italien* (Paris, 1743), 568. Cf. with Annibale Antonini, *Dictionaire italien, latin, et françois* (Paris, 1735), 691. The Cartesian example is again presented in *Dictionnaire italien, latin et françois: un abrégé du dictionnaire della Crusca*, 1755, 524, but is absent in the posthumous *Dizionario italiano, latino e francese* (Lyon, 1770), *passim*. Antonini's definition is modelled on the entry VORTEX taken from Ephraïm Chambers, *Cyclopaedia, Or a Dictionary of Arts and Sciences*, vol. II (London, 1728).

38 *Ragionamenti*, V.

FIGURE 2.2 The engraving by Charles-Nicolas Cochin (1715–1790), completed in December 1743 following the drawing by Charles Natoire (1700–1777)

vein, Fontenelle's original interpretation of two stanzas from *Orlando Furioso* are in the Italian version taken verbatim from Antonini's own new edition.[39] Finally, the clear focus on literary and linguistic aspects is again confirmed in the Preface, where Antonini argues for the greater purity of the Italian language learned by foreigners compared to that of native speakers.[40]

Given this context, should we accept Antonini's word, and believe there is a virtuous noblewoman behind the *Ragionamenti*? A straightforward answer would be that attributing the translation of certain texts to an anonymous third party constituted a common ploy in the eighteenth century. In this case, the fiction would have served the dual purpose of strengthening the parallel with the bright marquise of the *Entretiens* and of making a case for Antonini's teaching abilities. After all, Antonini was also active as a translator: in 1728, he published a French translation of his friend and colleague Paolo Rolli's *Remarks upon M. Voltaire's Essay on the Epick Poetry of the European Nations* (1727). Its aim, as we can expect by now, was to champion Italian literature against Voltaire's attacks, during the long-lasting controversy on literary traditions which was agitating learned circles in Paris at the time.[41] However, two additional clues are worth mentioning here. First of all, Antonini's dedications to several prominent figures of the French aristocracy – from Adelaïde d'Harcourt to Louise-Henriette-Gabrielle de Lorraine – definitely leave some room for speculation.[42] Secondly, from 1742 Antonini's circle of acquaintances included the prominent scientist and translator of Newton's *Principia* Émilie du Châtelet. A close friend of Fontenelle, the marquise du Châtelet was learning Italian in 1735, although the name of her teacher remains unknown.[43]

---

39 Cf. *Ragionamenti*, 77 with *Orlando furioso*, vol. 3, 414–415: Fontenelle uses 'ma Belle', while Antonini chooses 'Madonna'.

40 *Rargionamenti*, VII–IX.

41 Paolo Rolli, *Examen de l'essai de M. de Voltaire sur la poësie épique* (Paris, 1728). For context, see Folena, *L'italiano in Europa*, 429–431.

42 See for instance the frontpages and prefaces from Tasso, *Gierusalemme liberata* and *Grammaire italienne à l'usage des Dames*.

43 Antonini was given a mandate by the Neapolitan Alfonso Carafa della Spina, duke of Montenero, to negotiate the marriage between the latter and Gabrielle-Pauline du Châtelet, one of the daughters of Émilie. Châtelet later employed Antonini to deliver a letter to the mathematician François Jacquier in Rome. See Ulla Kölving and Andrew Brown, eds, *La Correspondance d'Émilie Du Châtelet*, 2 vols. (Ferney-Voltaire: Centre international d'étude du XVIIIe siècle 2018), 31–32, 187–188, 395. On Châtelet and Fontenelle, see Eszter Kovács, 'Fontenelle and Emilie du Châtelet, Mediating Scholars and Philosophers', *La Lettre Clandestine* 28 (June 2020): 299–312. On the links between Châtelet and Italian circles, see Ruth Edith Hagengruber, ed., *Époque Émilienne: Philosophy and Science in the Age of Émilie Du Châtelet (1706–1749)* (Heidelberg: Springer, 2022).

One last aspect to consider here concerns the physical features of the *Ragionamenti*, which provide us with additional information. In the 1740s, Antonini started collaborating with Laurent-François Prault, a Parisian printer who initiated a whole series of Italian classics that listed Antonini's editions of Tasso and Ariosto, together with a collection of poems.[44] When confronting the frontispiece of the *Ragionamenti* with that of Prault's series, the volumes present exactly the same types and designs in both the ornamental frame and the decoration at its centre. The similarity also applies to the engravings inside the books, all executed in the same period by Charles-Nicolas Cochin, a long-time collaborator of Prault.[45] These striking correspondences seem to suggest that the mention of Brunet on the title page of the *Ragionamenti* could have been an attempt to exploit the popularity of the new edition of the *Entretiens*, while the translation, actually printed by Prault, was regarded by Antonini as a work of Italian literature in its own terms. In light of the above, rather than disseminating Cartesian physics for an Italian audience in Paris or elsewhere, Antonini's intentions in publishing the *Ragionamenti* should therefore be contextualised with his constant promotion of the Italian language. To Antonini, the *Entretiens* primarily constituted a literary text, and their translation fits with his own objectives as an author and lexicographer.[46]

---

44 *Raccolta di rime italiane* (Paris, 1744). Prault also issued Antonini's *Grammaire italienne pratique et raisonnée* (Paris, 1746) and the 1743 *Dictionnaire françois, latin, et italien.*

45 Christian Michel, *Charles-Nicolas Cochin et le livre illustré au XVIIIe siècle: avec un catalogue raisonné des livres illustrés par Cochin, 1735–1759* (Geneva: Librairie Droz, 1987), 36, 56, 68, 208, 216, 228.

46 A rather fascinating yet unsupported hypothesis on the *Ragionamenti*'s place of publication involves the clandestine printing press of Raimondo di Sangro, prince of Sansevero, which was active in Naples at the end of the 1740s. In Paris, Antonini was connected to Andrew Ramsay and Antonio Conti, both part of the international freemason network. When he came back to Naples in 1750, Antonini used Sansevero's press to publish Ramsay's *Viaggi di Ciro*, in an edition whose types, patterns and engravings by Cochin match the *Ragionamenti*'s. Another censored publication issued from Sansevero's press used a false address and attributed the translation to an Italian noblewoman: see Henri de Montfaucon de Villars, *Il conte di Gabali', ovvero ragionamenti sulle scienze segrete* (London [i.e. Naples], 1751). For context, Rosanna Cioffi Martinelli, *La Cappella Sansevero: arte barocca e ideologia massonica* (Salerno: Edizioni 10/17, 1987), esp. 84–87; Giuseppe Giarrizzo, *Massoneria e illuminismo nell'Europa del Settecento* (Venice: Marsilio, 1994); Antonio Trampus, *La massoneria nell'età moderna* (Rome: Laterza, 2011). On Conti's early interest in Fontenelle's *Entretiens*, see Guido Baldassarri, Silvia Contarini, and Francesca Fedi, eds., *Antonio Conti: uno scienziato nella République des lettres* (Padua: Il poligrafo, 2009).

Bernardino Vestrini's *Trattenimenti* (1751) between Galilean Science and Metaphysics

Soon after Antonini's Parisian edition, two other translations of Fontenelle's *Entretiens* appeared in Italy. In 1749, the Venice printer Francesco Pitteri published a collection of Fontenelle's works in three volumes. In the Preface, the anonymous translator tells us that the source for his *Trattenimenti su la pluralità dei mondi* is the Hague edition, 'che è la migliore' ('which is the best one').[47] The reference is to the *Œuvres diverses* (1728–1729), of which, together with the *Entretiens*, Pitteri only issued the most commercial ones: *Histoire des oracles* (1687) and *Lettres galantes* (1685). An ironic reviewer of this new adaptation pointed out how both the translator and Fontenelle defended themselves in advance against criticisms of the contents of the book, and how the latter's refusal to explicitly identify the moon's inhabitants with human beings was a fig leaf to an insightful audience. However, this prudence guaranteed Pitteri official authorisations to print the *Trattenimenti*, which were released at the end of 1748 by both the Inquisitor of the Holy Office in Venice and the Public Auditor for the Republic Angelo Calogerà. Showing Paris as the place of publication on the front page should therefore be considered as part of a commercial strategy, rather than a manoeuvre to print the volumes clandestinely. A prominent member of the republic of letters, Calogerà had revised several manuscripts for Pitteri and could well be the translator, as his Italian versions of Fénelon and Pierre-François Guyot Desfontaines had appeared in Venice only a few years earlier.[48]

The translator whose name we can indubitably associate with a self-standing Italian edition of the *Entretiens* is the cleric and polymath Bernardino Vestrini. Published in Arezzo by the official press of the local bishop's typographer Michele Bellotti, Vestrini's *Trattenimenti su' la pluralità de' mondi* came out in 1751. The translation had been originally completed privately in Cortona for the

47 BernardLe Bovier de Fontenelle, *Opere del signor Bernardo Fontenelle*, vol. 1 (Parigi [Venice], 1749), vii.

48 See François de Salignac de La Mothe-Fénelon, *Le avventure di Telemaco figliuolo d'Ulisse* (Venice, 1744); Pierre-François Guyot Desfontaines, *Nuovo Gulliver o sia viaggio di Giovanni Gulliver figliuolo del capitano Gulliver* (Venice, 1731); *Novelle della repubblica letteraria* (Venice, 1750), 73–74. For context, see Mario Infelise, 'Editoria e società a Venezia nel Settecento', in *Cultura, intellettuali e circolazione delle idee nel '700*, ed. Renato Pasta (Milan: FrancoAngeli, 1990), 180–186; Patrizia Bravetti and Orfea Granzotto, eds., *False date: repertorio delle licenze di stampa veneziane con falso luogo di edizione, 1740–1797* (Florence: Florence University Press, 2008), 76; Cesare De Michelis, *Settecento veneziano* (Venice: Marsilio, 2022).

noblewoman Teresa Tommasi Vagnucci a few years before, and the publication appears to have been Bellotti's idea. In fact, Bellotti wrote both the dedication to his patron Giovanni Brichieri Colombi – a jurist who was appointed supervisor of the tribunals for the Grand Duchy of Tuscany in 1746 – and the preface. In the latter, Bellotti confirms he possessed Antonini's edition and listed the alterations to the 'Traduzione Parigina' (Parisian Translation).[49] Instead of the 1742 *Entretiens*, Vestrini took the 1698 edition as his source text, and did not consider changing or adding what Fontenelle had kept about 'le varie distanze, e grandezze de' Pianeti' ('the different distances and sizes of the Planets'),[50] since contemporary astronomers still disagreed about them anyway. Vestrini's version received much praise both in Italy and in France for its better quality compared to Antonini's and enjoyed 'un grandissimo spaccio' ('a huge sale to the public').[51]

Unknown to present historiography, Bernardino Vestrini was born in Dicomano, near Florence, in 1715. After completing his early education under the Jesuits, he took religious orders in 1732, when he became a cleric regular of the order of the Scolopi or Piarists; for most of his life he devoted himself to his order's mission of educating the young. Vestrini soon moved to Cortona as professor of moral philosophy and theology at the local seminary, and by 1755 had obtained the prestigious position of lecturer in Theology at the Collegio Nazareno in Rome. His scientific interests are already evident in his *Lettere teologiche* (1749–1751), a collection of reflections on different subjects. Besides traditional topics pertaining to the genre – like the immortality of the soul or the existence of God – Vestrini also indulged in natural philosophy, drawing for instance conciliatory parallels between Pierre Gassendi's atomism and the Holy Scriptures on the origin and matter of the Earth.[52] Later on, Vestrini published a classical historical dissertation on Lake Trasimeno (1756) and the *Relazione sullo stato del fiume Bigurro* (1767), showing a remarkable knowledge

---

49 Bernard Le Bovier de Fontenelle, *Trattenimenti su' la pluralità de' mondi di monsieur de Fontenelle trasportati dalla lingua francese nell'italiana da un accademico di Cortona* (Arezzo, 1751), VIII.

50 *Trattenimenti*, VIII.

51 *Storia letteraria d'Italia*, vol. 3 (Venice, 1752), 383; *Le journal des sçavans* (Amsterdam, May 1752), 403–404; Antonello Sbietti, 'Michele Bellotti (1749–1778) e l'edizione aretina delle opere di Ludovico Antonio Muratori (1767–1780)' (PhD diss., University of Siena, 1998).

52 Gassendi's works had been reprinted in Florence with a preface constituting a manifesto to promote empiricist science within the context of Italian Catholicism. See *Petri Gassendi Opera omnia* (Florence, 1727); Marcello Verga, 'La Cultura del Settecento. Dai Medici ai Lorena', in Furio Diaz, ed., *Storia della civiltà toscana, vol. v, I Lumi del Settecento* (Florence, 1999): 125–152.

of hydraulics which he applied to the improvement of the Grand Duchy's territories. Vestrini's study of fluid dynamics was indebted to the 'mai a bastanza lodato Galileo' ('never sufficiently praised Galileo') and his circle of physicists Famiano Michelini, Benedetto Castelli and Raffaello Staccoli.[53]

When Vestrini retired to the Tuscan countryside towards the end of his life, he conducted a long correspondence with his fellow academician Gian Girolamo Sernini. Their conversations revolved around ancient inscriptions, Italian politics and the reforms of Peter Leopold's administration, which Vestrini endorsed.[54] It had been in Cortona that Vestrini's social and intellectual activities had developed, when in April 1748 the *Accademia Etrusca delle antichità e delle iscrizzioni* (Etruscan Academy of antiquities and inscriptions) accepted him as a member. Founded in 1728 for the study of Etruscan civilization by the Venuti brothers Filippo, Marcello and Ridolfino, the Academy soon grew into a pivotal centre for Italian cultural and scientific renovation, and combined a traditional structure founded on the local erudite nobility with an open outlook on European novelties. Vestrini donated to the expanding Academy's library some manuscript codices containing works by the fifteenth-century humanists Leonardo Bruni and Francesco Filelfo. The library likewise conserved the academic dissertations publicly discussed at the Academy by its members, which alongside Vestrini's own ones contained speeches by Ludovico Muratori, Scipione Maffei and Giovanni Poleni.[55] In 1753, Vestrini was elected to the committee aimed at reforming the Academy's rules to keep up with its European counter-

53 Bernardino Vestrini, *Relazione dello stato del fiume Bigurro e lavori in esso necessarj esibita a signori rappresentanti del consiglietto di Castiglion Fiorentino dal padre Bernardino Vestrini cherico regolare delle Scuole Pie* (Arezzo?, 1767), 15; *Novelle letterarie pubblicate in Firenze* (Florence, 1786), 355–356; Bernardino Vestrini, 'Lettera VIII. Dell'Origine, Disposizione, Materia e Figura del Mondo', in *Lettere teologiche del padre Bernardino Vestrini delle scuole pie*, vol. 1 (Arezzo, 1749), 87–89; Bernardino Vestrini, *Dissertazione di un accademico Etrusco sopra l'emissario del lago Trasimeno arrichita di varie note, e di monumenti riguardanti la storia de' bassi tempi* (Rome, 1756).

54 Biblioteca del Comune e dell'Accademia Etrusca, Cortona, hereafter BCAEC, MS 675, c. 115^{r-v}, 120–122, 126–138, 148, 183, 242. Vestrini developed a further interest in political economy, see his translation of Victor de Riqueti, Marquis de Mirabeau, *La scienza cioè i diritti e i doveri dell'uomo* (Florence, 1774) and the historical treaty on the different Tuscan units of land measurements by Bernardino Vestrini, *Osservazioni sopra il braccio agrario fiorentino detto braccio a terra* (Florence, 1781).

55 Vestrini kept the records of the acts and proceedings of the debates for the Academy, participated in the study of local antiquities and won a poetry contest in 1750. See BCAEC, MS 438, 162–163, 199–200; MS 439, 89–92; MS 440, 72; MS 441, 26, 147; BMF, B VIII 6, 381^{r}–383^{v}, Bernardino Vestrini to Anton Francesco Gori, 12 April and 23 March 1752; B VIII 7, 304^{r-v} and 329, Marcello Venuti to Anton Francesco Gori, 27 May and 7 June 1750; Anton Francesco Gori, *Florilegium ex volumine VI. Noctium Corythanarum* (Florence, 1751), 26, 47.

parts. To a certain extent, this turned out to be necessary when Filippo Venuti promoted a new scientific branch of the Academy, now discussing the mercury thermometer, the microscope and the properties of synthetic phosphorus as well as archaeological findings.[56]

This innovative direction involved the translation of French and English texts, together with the repetition of the experiments they described. In 1748, at the house of abbot Filippo Alticozzi in Cortona, Vestrini witnessed the effects of the 'Macchina Elettrica' ('Electric Machine') as they had been observed by Jean Théophile Désaguliers in his *Dissertation sur l'électricité des corps* (1742).[57] For the occasion, Sernini produced an Italian translation of Désaguliers's text, to which Vestrini reacted in the form of a letter to Marcello Venuti. The letter is interesting for two main reasons. On the one hand, Vestrini comments on the translation practice itself: for him, 'il Traduttore è servo, non Padrone: trascrive, non compone' (the Translator is a servant, not a Master: he transcribes, he does not compose),[58] and should be especially faithful to the vocabulary and language of the source text when it comes to scientific subjects. On the other, Vestrini invites Sernini to consider more carefully his own reasons for undertaking the translation. Instead of conveying the experimental character of Desaguliers's working hypotheses, in the preface Sernini is eager to call for 'un nuovo Sistema da fondarsi sull'Elettricità' ('a new System to be founded on Electricity') that will shatter all the others.[59] In a rapidly changing scientific landscape, Vestrini warns, jumping to conclusions is a common mistake, and modest philosophers tend to pursue what is fashionable, replacing one system with another only in order to arouse new interest. To clarify his point further, Vestrini retraces the history of physics from Aristotle onwards:

---

56 See Bernardino Vestrini, 'Dissertazione VII, sopra l'uso sacro e profano degli agnelli', in *Saggi di dissertazioni accademiche: pubblicamente lette nella nobile Accademia etrusca, dell' antichissima città di Cortona*, vol. 5 (Rome, 1751), 135–156 and 'Dissertazione IV, sopra l'antica iscrizione ritrovata nel territorio di Asina Lunga' in *Saggi*, vol. 7 (Rome, 1758), 49–65; BMF, B VIII 7 v. 33, c. 442, Filippo Venuti to Anton Francesco Gori, 15 August 1753; Giammaria Mazzuchelli, *Gli scrittori d'Italia, cioè notizie storiche, e critiche intorno alle vite, e agli scritti dei letterati italiani* (Brescia, 1762), 1636; Girolamo Mancini, *I manoscritti della libreria del comune e dell'Accademia etrusca di Cortona* (Cortona: Stamperia Bimbi, 1884), 76–78; Paola Barocchi and Daniela Gallo, eds, *L'Accademia etrusca* (Milan: Electa, 1985).

57 Bernardino Vestrini, 'Lettera familiare intorno alla precedente Traduzione, ed alla Elettricità', in *Dissertazioni e lettere scritte sopra varie materie da diversi illustri autori viventi* (Florence, 1749), 53.

58 Vestrini, 'Lettera familiare', 44. The expression is borrowed from Pierre-François Guyot Desfontaines, *L'esprit de l'abbé Des Fontaines, ou reflexions sur differens genres de science et de litterature*, vol. 1 (London 1757), 304–305.

59 Vestrini, 'Lettera familiare', 49.

> Le figure, ed altre proprietà degli Atomi si suppongono, ma non si veggono; e ciò, che di essi conviene ammettere, non resta escluso dal nuovo sistema. I Vortici di Cartesio con i suoi Elementi s'immaginano, ma non si provano. L'Attrazione si nomina, e si suppone; ma fuori della calamita non si dimostra [...]. I calcoli Neutoniani principiano a non essere tanto rispettati quanto bisogna per la conservazione del suo sistema.
>
> (We can suppose the shapes, and other properties of the Atoms, but we cannot see them; and what we can accept about them is not excluded from the new system. We can imagine Descartes's Vortices with their Elements, but we cannot prove them. We can name Attraction, and suppose it; but cannot demonstrate it apart from the magnet [...]. The Newtonian calculations are beginning to be not as reliable as they need to be for the preservation of his system.)[60]

Hence, neither Cartesian vortices nor Newton's attraction can be demonstrated to be the universe's operating principles, and Vestrini gives us a simple reason for his belief: endorsing the Galilean process of *esperienze sensate* (sensible experiences), Vestrini believes we simply cannot observe and gather enough evidence on the *vortici* (vortices) and that Newton's mathematical method combined with his observations has become questionable. Therefore, the deductions of scientists are often 'piuttosto effetti de i nostri desideri, che delle vere leggi della Natura' ('the consequences rather of our desires than of the true laws of Nature').[61] If we take a closer look at the *Lettere teologiche* – published when the translation of the *Entretiens* was underway – Vestrini had already argued that the properties of planets could not be entirely perceived by us. In several passages, he eventually denies the ontological foundations of the Cartesian and Newtonian universal systems, rather displaying a phenomenological empiricism of clear Galilean tradition that had always resisted the theory of imperceptible vortices of matter composing the universe.[62] In the Grand Duchy, such a tradition had never been interrupted: while a censored version of Galileo's complete works was published in Florence in 1718, after

60 Vestrini, 'Lettera familiare', 51.

61 Vestrini, 'Lettera familiare', 52.

62 Paolo Casini, 'Il momento Newtoniano in Italia: un "post-scriptum"', *Rivista di storia della filosofia* 61, no. 2 (2006): 299–316. On the different ways in which Galileo's legacy was appropriated and developed in the eighteenth-century Italy, see Vincenzo Ferrone, *Scienza, Natura, Religione: Mondo Newtoniano e Cultura Italiana Nel Primo Settecento* (Naples: Jovene, 1982), 125–128.

the death of Cosimo III in 1737 Galilean experimentalism blossomed again in many of the environments now welcoming the ideas of Newton, Gassendi and Descartes, such as the University of Pisa, the Florentine court or other Tuscan academies.[63]

If Vestrini's 'agnostic' position in the clash between Descartes and Newton also resulted from prudence, his faith in the scientific method was unwavering even when addressing the sensitive subject of the plurality of inhabited worlds. Replacing Fontenelle's *Entretiens* with Christiaan Huygens's *Cosmotheoros* (1698), Vestrini ironically mentions how the origin of the world described in *Genesis* leads us to suppose that the Earth constitutes the centre of the universe, but 'basta supporre il Sistema Copernicano, ed Ugeniano per discredersi della nostra vanità' ('it is sufficient to suppose the Systems of Copernicus and Huygens in order to disbelieve our vanity').[64] From this perspective, continues Vestrini nonchalantly, the inhabitants of other planets would probably fall prey to the same error. The Holy Scriptures seem to assert 'che la Terra sia stabile, e permanente, e che da un sol Uomo sieno Venuti tutti gli altri' ('that the Earth is stable, and fixed, and that from one Man alone were descended all the others'),[65] thus rejecting today the hypotheses of Huygens and Copernicus. Yet, just as happened with the doctrine of the Antipodes, 'in avvenire con certe esperienze' ('in the future with certain experiments') these systems could be scientifically demonstrated, and the censorship of the Catholic Church eventually lifted.[66]

At the same time, the *Lettere teologiche* likewise affirm that God is the creator of the universe. Only fanatics like Epicurus, Vestrini tells his audience, could believe that in its wholesome perfection the universe is the result of a 'fortuito accozzamento d'insensati corpicciuoli' ('fortuitous meeting of tiny insensitive

---

63 See Vestrini, 'Lettera IX. Della Terra, e sua Sommersione nel Diluvio', 99; 'Lettera VIII', 93–94 and 'Lettera XI. Di Varie Produzioni, e Mutazioni delle Cose Naturali', 130, in *Lettere teologiche*, vol. I. For context, Francesca Bregoli, *Mediterranean Enlightenment: Livornese Jews, Tuscan Culture, and Eighteenth-Century Reform* (Stanford University Press, 2014), 176–181; Corey Tazzara, Paula Findlen, and Jacob Soll, eds., *Florence After the Medici: Tuscan Enlightenment, 1737–1790* (Routledge, 2019). A more complete edition of Galilei's dialogues, which included the censored *Dialogo sui massimi sistemi*, was issued in 1744 and is mentioned in Bellotti's preface to Vestrini's *Trattenimenti*, XI. See Galileo Galilei, *Dialogo dove ne i congressi di quattro giornate si discorre sopra i due massimi sistemi del mondo Tolemaico, e Copernicano* (Padua, 1744).

64 Vestrini, 'Lettera IX', 101. See Christiaan Huygens, *Kosmotheōros, sive de terris coelestibus earumque ornatu coniecturae* (The Hague, 1698), which also supported the hypothesis of a plurality of inhabited worlds.

65 Vestrini, 'Lettera VIII', 90.

66 Vestrini, 'Lettera VIII', 92.

bodies').[67] Metaphysicians extended this position further, arguing that beings do not exist because of a principle intrinsic to their nature, but have been created by God acting as a 'Causa estrinseca' ('extrinsic Cause').[68] God is at the origin 'anche del Moto, che egualmente è contingente nella materia, che ha bisogno di altra Causa, che la muova' ('of Motion as well, which is equally contingent in matter, which needs another Cause to move it').[69] For Vestrini, God gave the universe operational laws issued from his absolute wisdom, and gifted human beings with the faculties to understand these laws.[70] Therefore, it is necessary to investigate the physical world in order to 'adattare il senso incerto oscuro, e però versatile de i Sagri Libri, a quel vero invariabile, che si scuopre nelle leggi, ed opera della natura' ('adapt the uncertain, obscure, yet versatile meaning of the Sacred Books to the true invariable one that we discover in the laws and works of nature').[71] When paired with these metaphysical speculations, Vestrini's Galilean experimentalism seems to place him within a deterministic version of occasionalism similar to that expressed toward the end of his life by Malebranche, in which God acts in terms of intelligible general laws that manifest themselves in the natural world through apparent causality. This position remained consistent in Vestrini's approach to the *Entretiens* and was confirmed in his afterword.

In the first place, with knowledge of Vestrini's views on translation practices it emerges that he applied the advice given to Sernini to his own *Trattenimenti*. Vestrini's adaptation is meticulous in respecting the original text, and includes all of the contents of the *Entretiens*, regardless of their scientific validity. In this respect, Vestrini also remained faithful to Fontenelle's approach, or at least to that of his alter ego in the dialogues: in line with the sceptical tradition of seventeenth-century libertinage, the *Entretiens*'s courteous *philosophe* does not seek to prove the superiority of one system over the other, but discusses theories as hypotheses. Being a man of science, Vestrini adopted the same approach: his own considerations quoted above on the vanity of a system placing the Earth at its centre, for instance, are taken directly from the *Entretiens*, and

67 Vestrini, 'Lettera I. Dell'Esistenza di Dio', 3.

68 Vestrini, 'Lettera I', 4.

69 Vestrini, 'Lettera I', 5.

70 Vestrini, 'Lettera I', 8–9.

71 Vestrini, 'Lettera VIII', 93. Vestrini repeats this position several times. See also 'Lettera X. Dell'Antichità, e Durazione del Mondo', 127–128; 'Lettera XI', 129ff. As we can expect, Vestrini's ambiguous and unorthodox position sparked suspicion in Cortona, see Lorenzo Guazzesi, *Lettera del cavaliere Lorenzo Guazzesi aretino scritta al padre Bernardino Vestrini* (Arezzo, 1755), 3–4.

reproduced in the *Trattenimenti*.[72] Accordingly, Vestrini examined Fontenelle's text primarily as a scientific work, and it is safe to assume he carefully chose which edition to translate: as illustrated above, Bellotti's preface underlined the many measurements in the 1698 edition, some of which were excluded from the 1742 version chosen by Antonini.

Moreover, in this framework Vestrini's translation is particularly interesting because of the paratextual device concluding the *Trattenimenti*, namely a *Lettera del Traduttore* to Teresa Tommasi Vagnucci, the noblewoman who had asked for the Italian adaptation. In the letter printed as an afterword, Vestrini refers to the works of actual astronomers to rectify Fontenelle's obsolete Cartesian cosmology. For Descartes, the inertia caused by the motion God originally imparted to the matter of the universe after dividing it explained the formation of the vortex system, but Vestrini questions the scientific viability of this 'modo della loro formazione primiera' ('method of their original formation').[73] With this apparently minor remark, Vestrini actually moves away from the most radical occasionalist reading of Descartes's *Principia Philosophiæ* (1644), which would imply the reluctant acceptance of the laws of inertial motion and of the communication of motion by impact as the only fundamental laws of physics. Rather, Vestrini chooses to consult Huygens's *Cosmotheoros* again, this time for its materialist mapping of the universe.

In particular, Vestrini illustrates how Huygens's observations exposed the *Entretiens*'s wrong estimates of the distances and sizes of planets, and how his accurate update of the vortex system did not compute with the type and velocity of motion Descartes had originally ascribed to them.[74] 'Keplero, ed altri Astronomi' ('Kepler, and other Astronomers'), Vestrini carries on, replaced the planets' circular motion with an elliptical one,[75] removing the Sun from the centre of their orbits. Finally, to justify the force behind the revolution of planets, 'non si è finora trovato meglio della sentenza del Neuton' ('we have not thus far found anything better than Newton's assertion'), that is the law of universal

---

72 See Bernard Le Bovier de Fontenelle, *Entretiens sur la pluralité des mondes. Nouvelle édition, augmentée d'un nouvel entretien* (Paris, 1698), 36, and *Trattenimenti*, 17 and 'Lettera IX', 101.

73 Bernardino Vestrini, 'Lettera del Traduttore alla Nobiliss. Signora Teresa Tommasi Vagnucci', in *Trattenimenti*, 140.

74 Vestrini, 'Lettera del Traduttore', 140–141. Huygens accepted the existence of Newtonian vacuum and developed a concept of centripetal force to improve Descartes's model, while Fontenelle adjusted his 1708 edition's measurements of Venus and the Earth according to Huygens' astronomical data, see Pierre Connes, *History of the Plurality of Worlds: The Myths of Extraterrestrials Through the Ages* (Cham: Springer, 2020), 279–352.

75 Vestrini, 'Lettera del Traduttore', 140.

gravitation.[76] In a little more than a page, Vestrini shows he has a remarkable understanding of the latest developments in physics and astronomy, while also engaging with Descartes's original model of planetary motion beyond Fontenelle's version of it.

Later Translations and the Evolution of the *Entretiens* in Italy

Soon after Bellotti published Vestrini's *Trattenimenti*, the work became the object of a collective reading during one of the *notti coritane* (Cortonese nights), the private and informal meetings organised to discuss the latest publications in the republic of letters. The proceedings of that night hosted by Marcello Venuti do not tell us much about the audience's reaction to Vestrini's translation or to Fontenelle's Cartesian model, but emphasise the many qualities of the *Dama* (Lady) Tommasi Vagnucci, which are only matched by her 'splendida conversazione' ('wonderful conversation').[77] Likewise, it was Fontenelle's marquise and the dialogical model of the *Entretiens* that would resonate in many literary works of the second half of the eighteenth century.[78] While Italian publications seldom debated the system of vortices any more, the possibility of a plurality of worlds had a lasting impact on the collective imagination, with several authors across the country now producing both engaging philosophical refutations of such a dangerous heresy and amused remarks.[79]

In a series of Italian editions later published in Naples, we can observe a definitive shift in the perception of the *Entretiens*, now considered as a piece of French literature by their translators. In 1765, the physician Vincenzo Garzia edited the complete works of Fontenelle, incorporating the Venetian translations of 1749 with his own adaptations of the *Éloges des Académiciens* and of

76 Vestrini, 'Lettera del Traduttore', 140.

77 BCAEC, MS 440, fol. 56–57.

78 See for instance Vincenzo Ricci, *Stanze in lode di Cocco dell'Agguindolato acc. ordito di Padova con una cicalata dell'Innaspato* (Padua, 1750), 3–5; Saverio Bettinelli, *Il mondo della luna. Poema eroico-comico* (Venice, 1754), XII–XIV.

79 Giovanni Cattaneo, *Lettere moderne nè più nè meno di quel che sono. Pare che la plutalità de' mondi del Fontenelle vi abbia data occasione* (Venice, 1752); Alfonso Niccolai, *Dissertazioni e lezioni di sacra scrittura pubblicate da Alfonso Niccolai della Compagnia di Gesù*, vol. I (Florence, 1757), 120–129; Giovanni Cadonici, *Confutazione teologico-fisica del sistema di Guglielmo Derham Inglese* (Brescia, 1760); Appiano Buonafede, *Ritratti poetici, storici e critici di varj moderni uomini di lettere* (Venice, 1760), 31; Antonio Valsecchi, *Dei fondamenti della religione e dei fonti dell'empieta libri tre* (Padua, 1767), 29.

the literary texts included in the 1758 *Œuvres complètes*.[80] In 1784, the professor of Greek Francesco Mazzarella Farao reissued the 1748 Parisian translation with the original French text in parallel 'per giovare agli studiosi d'entrambe le Lingue' ('to benefit scholars of both Languages').[81] A great-nephew of Antonini, Mazzarella Farao slightly modified the original preface, where he mentions the 1728 experiments on stellar aberration by James Bradley as conclusive evidence for heliocentrism but stops short from entering further scientific debates, admitting they are beyond the interests of the field of literature.[82] Eventually, the man of letters Giuseppe Maria Galanti reissued this latest bilingual version for his series *gabinetto letterario* (literary cabinet) in 1788, removing Antonini's preface altogether. Instead, Galanti added a very short note stating that, whereas Fontenelle's system is a correct development of Galileian and Copernican theories, 'è erroneo tuttavolta nell'ipotesi cartesiana de' vortici' ('it is however erroneous in its Cartesian hypothesis of vortices').[83]

Nearly a century after the publication of the first edition of the *Entretiens*, the trajectory of their Italian translations appears to be complete. If considered as vehicles of transmission of knowledge, they still remained faithful to the scientific subjects of the original text notwithstanding the censorship of the Catholic Church. From the linear perspective of the history of science, Copernican heliocentrism remained relevant, while the Cartesian cosmology of vortices, still discussed in the 1730s, had by the middle of the century definitely become a relic of the past. Nevertheless, Fontenelle's masterpiece retained its significance as a literary work, and continued to be translated for its charming dialogues. In this sense, the careful disclaimer in Vestrini's *Trattenimenti* to regard Fontenelle 'come Poeta che finge, non come Filosofo che argomenta' ('as a Poet who imagines, not as a Philosopher who argues'),[84] initially shared in one form or another by most editors in order to elude the Inquisition, assumed its additional literal meaning.

80 Bernard Le Bovier de Fontenelle, *Opere del signor Bernardo di Fontenelle membro dell'Accademia francese*, 8 vols. (Naples, 1765).

81 Bernard Le Bovier de Fontenelle, *Ragionamenti su la pluralita de' mondi del sig. De Fontanelle tradotti dal francese in italiano da Francesco Mazzarella Farao* (Naples, 1784), 5.

82 Fontenelle, *Ragionamenti*, 10.

83 Bernard Le Bovier de Fontenelle, *Ragionamenti sulla pluralita de' mondi di M. De Fontanelle dell'Accademia francese con una nuova traduzione italiana a fronte del testo francese* (Naples, 1788), Avvertimento. A few years earlier, Galanti had defined the *Entretiens* as a *romanzo filosofico* (philosophical novel), see Giuseppe Maria Galanti, *Osservazioni intorno à romanzi, alla morale e a' diversi generi di sentimento* (Naples, 1781), 97 and Domenica Falardo, 'Riflessioni settecentesche sul romanzo. Un inedito di Giuseppe Maria Galanti', *Forum Italicum* 49, no. 1 (2015): 198–221.

84 Fontenelle, *Trattenimenti*, x.

However, when we consider the translations that constitute the object of this study as sites of knowledge, it becomes clear that both Antonini and Vestrini supplemented the *Entretiens* with extra, decisive information, and that this information was directly connected to the geographical and cultural contexts in which they produced their translations. In Paris, Antonini's preface, stylistic choices and printing practice aimed at affirming the superiority of the Italian language over French, emphasising their differences and therefore circulating the notion of cultural diversity within the broader Parisian context of the controversies involving European literatures. Antonini's 1748 *Ragionamenti* are essentially arranged so as to systematise Italian as the language of courteous conversation par excellence rather than to endorse or circulate Cartesian physics. On the contrary, in the academic environment of Cortona, Vestrini genuinely engaged with the cosmological theories Fontenelle had illustrated, although contradicting them through the afterword added to the 1751 *Trattenimenti*. This contradiction did not result from a displacement of meaning in the translation of the source text itself, but still led to the creation of an inherently conflictual site of knowledge: not so much to support and circulate the latest Newtonian beliefs, but to dismiss the Cartesian model of vortices altogether in the light of the Galilean scientific tradition Vestrini was part of.

Finally, in promoting their agendas as authors in their own right, both translators endorsed Fontenelle's initial intentions in different ways, and here again the local context plays a pivotal role in defining these endorsements. To Antonini, the dialogical model of the *Entretiens* proved appropriate for his lifelong commercial strategy of creating the popular genre of grammars 'for the ladies' of the Parisian salons, and fits with Fontenelle's purpose of introducing astronomy to a novice cosmopolitan readership. On the other hand, in the Grand Duchy of Tuscany Vestrini's Galilean background led him to share the sceptical attitude of the philosopher of the *Entretiens* when formulating viable scientific hypotheses. While still expressing a position compatible with the Catholic church, Vestrini like Fontenelle also dismissed the extreme occasionalism of a certain Cartesian metaphysical tradition: not to defend a strictly mechanist reading of Descartes's system, but rather to search for convincing physical evidence in the light of the latest scientific observations of astronomers.

Acknowledgements

I am immensely grateful to Dr. Alviera Bussotti, Dr. Patrizia Rocchini, Dr. Rita De Tata, Dr. Sara Malfatti, Prof. Ann Thomson and Prof. Mario Infelise for all their invaluable archival help, knowledgeable suggestions and editorial assistance.

CHAPTER 3

'Un Eroico Traduttore'

Pietro Antoniutti, Translator and 'Cultural Mediator' between Venice and Europe (1780–1820)

Alessia Castagnino

> In creating a new act of communication out of a previously existing one, translators are inevitably acting under the pressure of their own social conditioning while at the same time trying to assist in the negotiation of meaning between the producer of the source-language text (ST) and the reader of the target-language text (TT), both of whom exist within their own, different social frameworks.[1]

Over the last four decades, translation practices have been at the centre of renewed interest among scholars of various disciplines, including historians. While developing some of the most recent theoretical and methodological foundations of Translation Studies and Translation History,[2] historians have gradually started to take a close interest in translations as cultural products and results of a process of cultural and intellectual – not only linguistic – negotiation. At the same time, they have acknowledged their importance as 'documents of singular value'[3] and sources to understand historical phenomena and

---

1 *Literary Translator Studies*, ed. Klaus Kaindl, Waltraud Kolb, Daniela Schlager (Amsterdam: John Benjamins 2021), 1. The present article is part of a comprehensive study on the Italian circulation of William Robertson's *Histories*. Some preliminary reflections on Antoniutti's activities as a translator can be found in Alessia Castagnino, 'La plume du traducteur. Le prêtre Pietro Antoniutti médiateur culturel entre Venise et la Grande Bretagne, années 1780–1820', in *Prendre la plume des Lumières aux Romantismes. Pratiques de l'écrit dans l'Europe de la fin de l'époque moderne*, ed. Matthieu Magne (Clermont-Ferrand: Presses Universitaires Blaise-Pascal, 2019), 53–66.

2 One of the earliest works to emphasise the need for historians to study translations, by paying attention to their cultural dimension, was *Cultural Translation in Early Modern Europe*, ed. Peter Burke, and Ronnie Po-chia Hsia (Cambridge: Cambridge University Press, 2006). On the importance of establishing a dialogue between translation studies and historical studies, see the works by Christopher Rundle, for example, 'Theories and Methodologies of Translation History: the Value of an Interdisciplinary Approach', *The Translator* 20, no. 1 (2014): 2–8.

3 *Traduzioni e circolazioni delle idee nella cultura europea tra '500 e '700*, ed. Girolamo Imbruglia, Rolando Minuti and Luisa Simonutti (Napoli: Bibliopolis, 2007), 2.

 | DOI:10.1163/9789004731806_005

investigate the ways in which ideas and texts were effectively disseminated and re-elaborated in new cultural, social and political contexts.

The eighteenth century has proven to be an ideal research field to analyse these dynamics. During that period, as argued by scholars such as Fania Oz-Salzberger, Stephanie Stockhorst or Jesús Astigarraga – to mention just a few names[4] – translations were an extraordinary vehicle for the circulation and reception of a large number of works and a tool for the formation of new scientific, political, economic, philosophical and historical languages, as well as a remarkable business for publishers and printers.[5] Attention to translation strategies has enhanced our knowledge of the reception of texts in Enlightenment Europe; it demonstrates the vivacity and dynamism of the exchanges that took place between different cultural contexts, including that of 'arrival', which does not receive the text passively but significantly changes it.[6] In addition, this approach has highlighted the need to define and develop a research methodology capable of examining the translations in all their richness and complexity.

Taking up the suggestions put forward by scholars such as Lawrence Venuti, Anthony Pym or Michel Espagne,[7] scholars have begun to pay more attention

4 Within a vast literature which has been significantly enriched in recent years by important contributions, see at least Fania Oz-Salzberger, *Translating the Enlightenment. Scottish Civic Discourse in Eighteenth-Century Germany* (Oxford: Clarendon Press, 1995), a pioneering book which made historians think in a new way about the importance of translations for Enlightenment studies; and 'The Enlightenment in Translation: Regional and European Aspects', *European Review of History* 13 (2006): 385–409; *Cultural Transfer through Translation. The Circulation of Enlightened Thought in Europe by Means of Translation*, ed. Stephanie Stockhorst (Amsterdam-New York: Rodopi, 2010); See also *Cultural Transfers: France and Britain in the long eighteenth century*, ed. Ann Thomson, Simon Burrows and Edmond Dziembowski (Oxford: Voltaire Foundation, 2010).

5 László Kontler defined translations as an important 'tool of Enlightenment cosmopolitanism': László Kontler, 'What is the (Historians') Enlightenment Today', *European Review of History* 13, no. 3 (2006): 357–371 (364). On the importance of translations as medium that 'triple l'offre de lecture' see Patrice Bret, 'Enrichir le magasin où l'on prend journellement. La presse savante et la traduction scientifique à la fin du 18ème siècle', in *Les journaux savants dans l'Europe moderne. Communication et construction des savoirs*, ed. Maria Conforti, Patrizia Delpiano, and Jeanne Peiffer, special issue of *Archives internationales d'histoire des sciences* 63 (2013): 359–382 (361).

6 See Fania Oz- Salzberger, 'Enlightenment, National Enlightenment, and Translation', in *The Routledge Companion to Eighteenth Century Philosophy*, ed. Aaron Garrett (London: Routledge, 2014), 31–61; Jesús Astigarraga, 'Les traductions espagnoles des normes législatives et des écrits économiques de Turgot (1774–1791)', *Annales historiques de la Révolution française* 386 (2016): 27–52.

7 Among the many works on the importance of translators, in addition to the seminal work of Lawrence Venuti, *The Translators Invisibility. A History of Translation* (London-New York:

not just to the so-called 'material vectors', but also to 'social' ones. While in the first case, interest is focused on the changes that a work may undergo in the translation process (from linguistic and stylistic adaptations to changes to the format, the text structure or the paratextual elements), in the second, the focus is on all those professionals (from translators, publishers and editors to the sponsors of the publishing projects and reviewers, who comment on and discuss such linguistic and cultural adaptation projects on the pages of literary journals and periodicals) who operate at various levels in the transfer of a work from the context of production to the target context. As Espagne has pointed out, all these figures, and in particular the translators themselves, do not merely make simple stylistic and linguistic adaptations; they actually act at different levels as mediators who are conditioned by 'their motivations, their stories, their economic dependence, and their ideas about what needs to be communicated in a new context in order to fill an intellectual gap'.[8] They have precise goals to achieve with their own specific work, which may be the result of a personal initiative or, on the contrary, influenced by editorial or business considerations. A translation can also be the tangible result of broader projects, such as those promoted and financed by governments, academies, or scientific and literary societies in order to circulate works of benefit to the public. Contacts established by translators with travellers, diplomats, or foreign men of letters may influence their choices as to what does or does not deserve to be translated. Or their travels, as in the case studied here, could lead them to visit libraries, academies, and discover works that they then decide to make available in their own language. Whatever the motivation behind the production of a given translation, there is always a specific work of mediation. As a result,

Routledge, 1995), see Anthony Pym, 'Humanizing Translation History', *Hermes-Journal of Language and Communication Studies* 42 (2009): 23–48; Jean Delisle, *Les traducteurs dans l'histoire*, 2nd ed. (Ottawa: Ottawa University Press, 2007); and Mirella Agorni, 'A Marginal(ized) Perspective on Translation History: Women and Translation in the Eighteenth Century', *Meta* 50, no. 3 (2005): 817–830. On the so-called 'sociological turn' in translation studies see Andrew Chesterman, 'The Name and Nature of Translator Studies', *Hermes-Journal of Language and Communication Studies* 42 (2009): 13–22.

8 '[I traduttori sono influenzati] dalle loro motivazioni, dalle loro storie, dalle loro dipendenze economiche e dalle loro idee su ciò che deve essere comunicato', in Michel Espagne, 'Il ruolo della traduzione nella genesi del Neoclassicismo', in *Traduzioni e traduttori nel Neoclassicismo*, ed. Giulia Cantarutti, Stefano Ferrari and Paola Maria Filippi (Milano: Franco Angeli, 2010), 13–22 (21). All translations of original Italian passages are mine. By the same author see also 'La fonction de la traduction dans les transferts culturels franco-allemands au XVIIIe et XIXe siècle: le problème des traducteurs germanophones', *Revue d'histoire littéraire de la France* 97, no. 3 (1997): 413–427.

a product is delivered into the hands of the readers that is never neutral but reshaped and readapted in both its textual and peritextual aspects.

This chapter constitutes a case study to test these methodological reflections and theoretical perspectives, by exploring the activities of one of these 'cultural mediators': the Friulian priest Pietro Antoniutti who, thanks above all to his outstanding career as a translator, managed to carve out a prominent position for himself in the Venetian cultural panorama of the late eighteenth century. In 1974, the Italian historian of philosophy Eugenio Garin described Antoniutti as a 'heroic' and uncommon translator, given the large number of translations that he produced as well as the fact that he often covered the printing costs himself; in a little over thirty years, the priest had more than thirty translations printed, while according to his nineteenth-century biographers a further fifteen translations remained in manuscript.[9] Each of these translations was the result of a meticulous adaptation of a rich selection of English and Scottish works that, as we will see, he considered particularly useful to provide his Venetian readers with new reflections and alternative viewpoints and to help them understand the changes taking place in the society and political institutions of the Republic of Venice *au tournant des Lumières*.

The main purpose of this chapter will be to explore Antoniutti's career in order to show how his own life experiences, travels and encounters influenced his in-depth knowledge of British culture and literature and, even more importantly, his own conception of the translator's craft and 'mission'. It begins with a brief analysis of a number of key moments in the priest's biography which throw light on his translation strategies. Then I will study in detail some of his most important and representative translations, following the broadening and diversification of his interest in British culture. The focus here will be on specific aspects of the process of translation, mainly the motivations behind each project and the way he adapted the original editions. The function of new forewords, letters of dedication and footnotes will be analysed in order to show the extent to which these peritextual elements transformed and reshaped the original author's ideas for the new reading audience. Specific attention will be paid to his translation of William Robertson's *History of Scotland*, a

9 Eugenio Garin, 'Ferguson in Italia', *Rivista critica di storia della filosofia* 4 (1974): 96–97. The definition 'eroico traduttore' was first used by Giulio Natali (ed.), *Storia della letteratura*, vol. 3: *Il Settecento* (Milano: Vallardi, 1962), 543. See also Paul Hazard, *La Rivoluzione francese e lettere italiane, 1789–1815*, trans. Pier Antonio Borgheggiani (Rome: Bulzoni, 1995 [Paris, 1910]), 475.

work that well sums up all the main features of his strategies of peritextual adaptation, which set him apart from other translators of Robertson's masterpiece.

Antoniutti's Career between Venice and Constantinople

Pietro Antoniutti was born on 25 November 1732 in San Daniele del Friuli, which had been ruled by the Republic of Venice since the first decades of the sixteenth century.[10] Before being ordained a priest and moving to Venice at the end of the 1740s, he studied at the Jesuit college of Gorizia; this lively cultural environment brought him into contact with some of the most up-to-date European publications and, at the same time, enabled him to start learning both classical and modern languages, first and foremost French and English. He pursued this inclination for linguistic studies and, more generally, his interest in what was being debated and theorised beyond the Italian peninsula throughout the 1750s and the 1760s; a number of professional appointments led him to spend a few months in Vienna before moving to Constantinople, where he took on the role of chaplain and confessor of Giovan Antonio Ruzzini, the *bailo* (ambassador) to the Ottoman *Sublime Porte*.

As we can read in the detailed obituary written by the Somascan Father Giannantonio Moschini and published in 1827 in the *Giornale sulle scienze e lettere delle Provincie Venete*, Antoniutti's stay in Vienna was of particular importance both for his career as a translator and for the maturing of his cultural interests; he was able to visit libraries and 'conversare con dotti uomini' (converse with erudite men), who encouraged him to 'applicarsi sommamente alla traduzione di ottimi libri dalla lingua inglese' ('fully commit himself to

10 No detailed study of Pietro Antoniutti's biography and career is currently available. Some biographical information and a partial list of his translations are provided by Attilio Zadro, 'Pietro Antoniutti e la consapevolezza storica nelle Venezie fra il XVIII e il XIX secolo', *Giornale critico della filosofia italiana* 69 (1988): 71–79, and Giovanni A. Moschini, 'Necrologia di Pietro Antoniutti'. *Giornale sulle scienze e lettere delle Province Venete* 12 (1827): 287–288; this article was published anonymously but can be attributed to the Somascan Father Moschini: see Emanuele A. Cicogna, *Saggio di bibliografia italiana* (Venice: Merlo, 1847), 375. See also Girolamo Dandolo, *La caduta della Repubblica di Venezia ed i suoi ultimi cinquant'anni* (Venice: Pietro Naratovich, 1855), esp. 213–214; Luigi Narducci, *Notizie biografiche del sac. Pietro Antoniutti* (Udine: Tipografia del Patronato, 1884); and Sante Della Valentina, 'Memorie de' Preti secolari Alunni delle Chiese parrocchiali di Venezia che si distinsero in fatto di scienze e di lettere nel secolo XVIII', ms. 2186, c. 364, cc. 359–362, collection Cicogna, Biblioteca Correr, Venice.

the translation of excellent books written in English'), like David Hume's *History of England*.[11] In the Ottoman capital, Antoniutti broadened and further diversified his knowledge of English and Scottish historiography, philosophy and literature, thanks to his close friendship with John Murray, British Ambassador to the Ottoman Empire, who invited him to stay on in Constantinople even after Ruzzini's departure. Murray's intellectual circle brought him into contact with diplomats, ambassadors, aristocrats and travellers on the Grand Tour, which increased his curiosity about multiple aspects of British culture; this also helped him to perfect the linguistic skills that would allow Antoniutti to distinguish himself in subsequent decades as a translator capable of working directly from the original editions of works written in English.[12]

In the early 1770s, Antoniutti decided to return to Italy in the company of Lord Edward Wortley Montagu, son of the famous Lady Mary Montagu.[13] He initially settled in Padua, where for some years he served as the private tutor of Count Valvasone's children, though with varying success. In 1789 he was appointed chaplain to the Doge Ludovico Manin, and on 23 July 1793 he became the parish priest of San Giacomo in Rialto, a position he would maintain until his death in March 1827.[14]

Thanks to the economic stability provided by these appointments, Antoniutti was able to devote himself almost full-time to his work as a translator, putting to good use the linguistic skills and knowledge acquired in previous

11 Moschini, 'Necrologia di Pietro Antoniutti', 288. Antoniutti reconstructed this episode in his *Introduzione alla Storia d'Inghilterra di David Hume* (Venice: Santini 1816), where he described Hume's historical work and announced to the readers the imminent publication of his Italian translation.

12 It was a common practice of Italian publishers to translate English works not from the original editions but from the French versions that were already available. The ability to translate directly from the original English editions was highly appreciated by both readers and publishers. The latter considered that it provided added value that would guide the choice of potential customers. On the phenomenon of 'second hand translation' and its cultural implications see Génèvieve Roche, *Les traductions-relais en Allemagne au XVIIIe siècle. Des lettres aux sciences* (Paris: CNRS Editions, 2001).

13 The friendly relationship between Antoniutti and Edward Montagu is mentioned by the priest himself in his forewords to Edward Wortley Montagu, *Riflessioni sopra la elevazione e la dicadenza delle antiche repubbliche adattate al presente stato della Gran Bretagna dal Cavalier Edoardo Montagu, tradotte dall'originale inglese* (Udine: Gallici, 1781), and to Elizabeth Montagu, *Saggio sopra gli scritti e 'l genio di Shakespeare paragonato a Greci e Francesi drammatici di madama Montagù, volgarizzamento di Pietro Antoniutti* (Venice: Palese, 1803).

14 *Gazzetta urbana veneta*, 2 May 1789, 277.

decades. As Father Moschini emphasised, from the 1780s onwards, Antoniutti 'spendeva tutto il suo tempo' ('spent all his time') translating the best British philosophical and historical works, and did not hesitate to invest his money in publishing his translations, which he then gave 'ai suoi amici' ('to his friends').[15] These were Venetian aristocrats, members of the Republic's ruling class whom he wanted to educate and bring up to date by making available the best works by British authors, which were so rich and exhaustive that they could also be applied to the Venetian context. These Venetian aristocrats frequently also featured as the recipients of the dedications of his publications (for example the Doge Lodovico Manin, but also aristocrats like Francesco Pesaro and Girolamo Ascanio Giustiniani, who would play key roles in the late eighteenth-century Venetian government).[16]

It is important to emphasise that Antoniutti's first translation projects originated and developed at a specific moment in the political and cultural history of the *Serenissima* Republic of Venice: as it faced a profound crisis on both the institutional and socio-economic levels, the majority of the Venetian government's members became aware of the urgent need for a thorough and systematic change in its administrative system and productive structures.[17] Thus, from the 1760s onwards, exponents of the ruling class started to devise and propose reforms capable of improving the Republic's institutional, economic and social system.

As a result, in this period, various publishing projects were promoted that aimed at circulating, among the broadest possible audience, not only the original works of Venetian writers and scientists, but also translations of works published in other European contexts that could stimulate reflection and discussion on how to renew the government and economy of the Venetian state. In the field of political economy, for example, the patrician Matteo Dandolo trans-

15 Moschini, 'Necrologia di Pietro Antoniutti', 288.

16 On the role of dedications see Marco Paoli, *L'appannato specchio. L'autore e l'editoria italiana nel Settecento* (Lucca: Pacini Fazzi, 2004), and Annalisa Nacinovich, 'Diplomatici e scienziati nei carteggi rediani', in *Diplomazia e comunicazione letteraria nel secolo XVIII: Gran Bretagna e Italia / Diplomacy and Literary Exchange: Great Britain and Italy in the long 18th Century*, ed. Francesca Fedi and Duccio Tongiorgi (Rome: Edizioni di Storia e Letteratura, 2017), 19–36.

17 See Franco Venturi *Settecento riformatore*, vol. 5: *L'Italia dei Lumi*, vol. 2: *La Repubblica di Venezia, 1761–1797* (Turin: Einaudi, 1990), and *Storia di Venezia dalle origini alla caduta della Serenissima*, vol. 8: *L'ultima fase della Serenissima*, ed. Piero Del Negro and Paolo Preto (Roma: Istituto dell'Enciclopedia Italiana, 1998). On the eighteenth-century publishing market in Venice see Mario Infelise, *L'editoria veneziana nel Settecento* (Milan: FrancoAngeli, 1989).

lated David Hume's *Political Essays*,[18] while printers such as Gasparo Storti, Giovanni Gatti or Antonio Zatta offered their readers Italian versions of works on the history of scientific, technical, philosophical or literary progress in different parts of Europe in the early modern era.[19] More generally, throughout the eighteenth century, translations were considered – not only in Venice – a fundamental tool for the circulation of knowledge that was 'useful' for the education and information of an increasingly broad public made up of experts as well as 'common' readers.

Antoniutti's translating activities fit perfectly in this context. Evidence of the priest's appreciation of the potential of translations to enhance the cultural growth of both educated readers and a more general public emerges from the forewords and letters of dedication he included in his main works.[20] There, he informed Italian readers of his 'mission', namely his intention to educate and inform the public about the most useful British innovations – philosophical and historiographical as well as a more specifically political and economic. His aim was to make good use of the linguistic skills and knowledge he had acquired thanks to the networks built up during his travels in Vienna, Constantinople and Venice. His intense translation activities were all characterised by a real autonomy in the choice of works to be translated, and by a precise idea of what could and should be the most appropriate way of carrying out these translations – except, of course, for any additions, omissions or corrections required by the censors at the time of publication.

---

18 Matteo Dandolo, *Saggi politici sopra il commercio del signor David Hume. Traduzione dall'inglese di Matteo Dandolo patrizio veneto* (Venice: Bassaglia e Pavini, 1767).

19 On the Italian and Venetian 'translation market' see Daniela Goldin Folena, 'Traduttori e traduzione in Europa e nel Veneto tra Sette e Ottocento', in *Vie lombarde e venete. Circolazione e trasformazione dei saperi letterari nel Sette-Ottocento fra l'Italia settentrionale e l'Europa transalpina*, ed. Furio Brugnolo, Helmut Meter (Boston: De Gruyter, 2011), 135–154; Alessia Castagnino, '*Il bisogno della traduzione*. Lecteurs, éditeurs et stratégies de traduction dans le long XVIIIe siècle italien', in *Lire en Europe. Textes, formes, lectures (XVIIIe–XXIe siècles)*, ed. Lodovica Braida, Brigitte Ouvry-Vial (Rennes: PUR, 2020), 111–128. On the translation of scientific texts in Venice, see the chapter by Giovanni Lista in this volume.

20 For more on the use of introductions, prefaces, and other paratextual elements – and on the importance of studying these elements attesting the translator's 'voice' – see Şehnaz Tahir Gürçağlar, 'The Translator as Subject: Literary Translator Biographies, Memoirs and Paratexts', in *The Routledge Handbook of Literary Translation*, ed. Kelly Washbourne, Ben Van Wyke (Abingdon: Routledge, 2018), 524–537.

From Political to Historiographical Works: Pietro Antoniutti's First Translations

In view of what I have said above, the first stages of Antoniutti's translation career hardly seem surprising. He initially focused his attention on Lord Montagu's *Reflections on the Rise and Fall of the Ancient Republics*; the translation was published in 1781 by the printer Gallici in the Friulian city of Udine. Although he decided to maintain his anonymity and not to mention his name on the volume, the reference in the Introduction to Antoniutti's friendship with Montagu together with references to this translation in others of his works inevitably revealed the translator's identity.[21]

The topics discussed in the *Reflections* – also mentioned explicitly in the title of the original edition – perfectly fit the mindset of the translator and his purpose to 'formare un vero cittadino della Repubblica' ('to form a true [conscious] citizen of the Republic of Venice').[22] From the beginning of his foreword, Antoniutti encouraged his readers to pay particular attention to Montagu's claims about the possible causes of the decline of the Greek and Roman republics' systems of government, and to his emphasis on the key role played by laws and legislators in ensuring all citizens 'prosperità e libertà' ('prosperity and freedom'). According to Antoniutti, this short but precise work ('ristretta, ma precisa esposizione'), provided excellent examples of systems of government 'da seguire ed emulare' ('to follow and emulate') in order to create optimal conditions for the growth of public welfare and well-being.[23] He invited his readers – and especially all those who had to 'fondare o riformare stati' ('to found or reform a state') – to read the Reflections carefully, drawing comparisons with what was happening in Venice in those same years, in order to find possible analogies.[24]

This publication was rapidly followed by a second translation, aimed at giving his reading audience access to some of the political and philosophical works of Benjamin Franklin, one of the few non-British authors he translated.[25]

21 Antoniutti made an explicit reference to his work in a letter addressed to Giuseppe Remondini (see *Epistolario Remondini*, Biblioteca Civica di Bassano).

22 Wortley Montagu, *Riflessioni sopra la elevazione e la dicadenza delle antiche repubbliche*, xv.

23 *Riflessioni*, vi–vii.

24 Ibid. Antoniutti also took the opportunity to develop his own reflection on the effectiveness of the 'costituzione Britannica' (British constitution), which referred to the fundamental principles of the Roman and Greek political systems.

25 Benjamin Franklin, *Opere politiche di Beniamino Franklin L.L.D.F.R.S. nuovamente raccolte e dall'originale inglese recate nella lingua italiana* (Padua: Manfrè, 1783); *Opere filosofiche*

This important project represented a new occasion to urge a more in-depth reflection on the crises and revolutions that may confront modern states and their governments.

According to various critics and, in particular, the anonymous author of an article published in the Venetian literary journal *Progressi dello spirito umano nelle scienze e nelle arti*, one of the translation's qualities was Antoniutti's ability to include footnotes that were always very accurate, 'molto erudite e utili, mai fuori posto o ridondanti' ('very erudite and useful, never out of place or redundant').[26] For instance, one of these footnotes contained a reference to the work by Edward Montagu that he had previously translated, aimed at connecting the ideas of the two authors and identifying similarities and differences between their works and reflections on the rise and fall of European political systems. The inclusion of references to other translations of his was a common technique, as if Antoniutti wanted to create an ideal thread connecting all his translations, thus giving readers a complete picture of the issues discussed and guiding them along a path of highly recommended texts that would be useful for their education.

Antoniutti's interest in British literary productions and political experiences was so strong that, in 1791, he decided to prepare an Italian version of Edmund Burke's *Reflections on the Revolution in France*. It had already been published once in Italian, by the printer Giacomo Storti who, however, had based his translation on an existing French version. Contrary to what one might think, Antoniutti did not choose this work because of its central focus on the recent events of the French Revolution, undoubtedly a popular theme at the time. His intention was far more complex. Burke's work was, in fact, merely a pretext to be able to 'offer his readers a direct and objective documentation' of the great debate on the English Constitution stimulated by the pressure of events in Paris.[27] In order to shift the readers' attention to the issues that interested

di Beniamino Franklin nuovamente raccolte, e dall'originale inglese recate in lingua italiana (Padua: Conzatti, 1783); *Osservazione a chiunque desideri passare in America; e riflessioni circa i selvaggi dell'America settentrionale del d.re Franklin dall'originale inglese recate in lingua italiana da Pietro Antoniutti* (Padua: Conzatti, 1785). On the circulation of Franklin's ideas and works in eighteenth-century Italy see Antonio Pace, *Benjamin Franklin and Italy* (Philadelphia: The American Philosophical Society, 1958); on interest in America more generally see Piero Del Negro, *Il mito americano nella Venezia del '700* (Padua: Liviana Editrice, 1986).

26 *Giornale letterario, ossia Progressi dello spirito umano nelle scienze e nelle arti*, February 18, no. 7 (1784): 191–195 (194).

27 '[Antoniutti voleva] offrire ai suoi lettori una documentazione diretta ed obiettiva del grande dibattito sulla costituzione inglese provocato dall'incalzare degli avvenimenti pari-

him most, Antoniutti intervened in the text, adding long passages taken from works by Richard Price, Catherine Macaulay Graham and Joseph Towers. As he explained in detail in his foreword, the 'Avvertimento al lettore', his aim was to present the ideas of authors who were less known in the peninsula, but who were crucial to understand the direction in which the European debates on democracy and sovereignty were moving. In other words, he once again sought to present alternative viewpoints that would be useful to understand how to tackle the challenge of keeping alive 'il dolce e libero governo di Venezia' ('the gentle and free government of Venice').[28] Not surprisingly, the priest chose to dedicate this Italian edition to Girolamo Ascanio Giustiniani, a leading member of the patrician ruling class in the second half of the eighteenth century who died while the translation was in print.[29]

One thing that all these examples have in common is the fact that the translator adopted a systematic strategy of textual and paratextual adaptation. This strategy did not imply any specific linguistic or stylistic interventions; each version remained as faithful as possible to the original, with the exception of some errors due, in all probability, to the fact that his fluency in English was, despite everything, still far from perfect. In this connection, it is interesting to quote the opinion of the Venetian priest and author Sante Della Valentina, who emphasised that Antoniutti was 'constantly itching to translate and publish his versions of works translated from the English language'; according to Della Valentina, Antoniutti's choices were the best possible ones, but the result of his translation left much to be desired, especially because of his poor knowledge of the Italian language.[30]

---

gini', in Roberto Zapperi, 'Burke in Italia', *Cahiers Vilfredo Pareto. Revue européenne d'histoire des sciences sociales* 3, no. 7/8 (1965): 5–62 (22).

28 'Avvertimento', in *Tre lettere apologetiche di Mrs Macaulay Graham, di Mr Towers, e del Dr Price contra le "Riflessioni" di Mr Edmund Burke, sopra la Rivoluzione di Francia con un breve compendio delle stesse; il tutto volgarizzato dagli originali Inglesi da Pietro Antoniutti* (Venice: Tosi, 1791), III.

29 See Giuseppe Gullino, 'Giustinian, Girolamo Ascanio', in *Dizionario Biografico degli Italiani* (Rome: Istituto dell'Enciclopedia Italiana 2001), vol. 57, https://www.treccani.it/enciclopedia/girolamo-ascanio-giustinian_(Dizionario-Biografico)/

30 'Pietro Antoniutti ha il continuo prurito di tradurre cose tradotte dall'inglese: la scelta che fa di esse è la migliore possibile, ma anzicché tradurle le tradisce con le sue versioni troppo servili e senza il sicuro possesso della lingua italiana', in Della Valentina *Memorie de' Preti*, 359. The *topos* of the necessity and importance of knowing both the source and target languages was well established in Italian treatises on translation. Antoniutti, however, was criticised for using an Italian that was too archaic and incapable of capturing the tastes of Venetian readers.

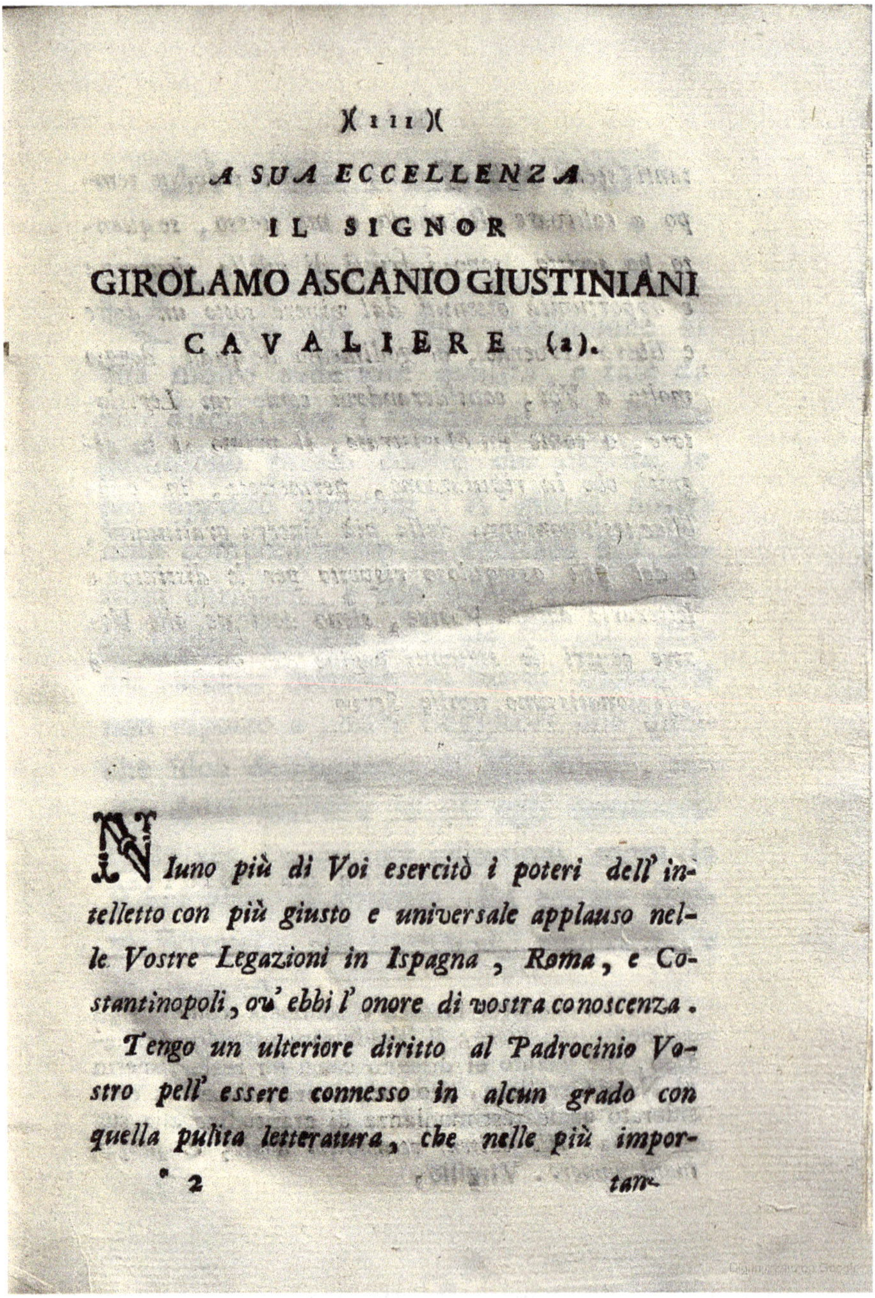

)(III)(

A SUA ECCELLENZA

IL SIGNOR

GIROLAMO ASCANIO GIUSTINIANI

CAVALIERE (a).

Niuno più di Voi esercitò i poteri dell' intelletto con più giusto e universale applauso nelle Vostre Legazioni in Ispagna, Roma, e Costantinopoli, ov' ebbi l' onore di vostra conoscenza.

Tengo un ulteriore diritto al Padrocinio Vostro pell' essere connesso in alcun grado con quella pulita letteratura, che nelle più impor-

* 2 *tan-*

FIGURE 3.1A Antoniutti's dedication, in *Tre lettere apologetiche* (Venice, 1791). Private collection

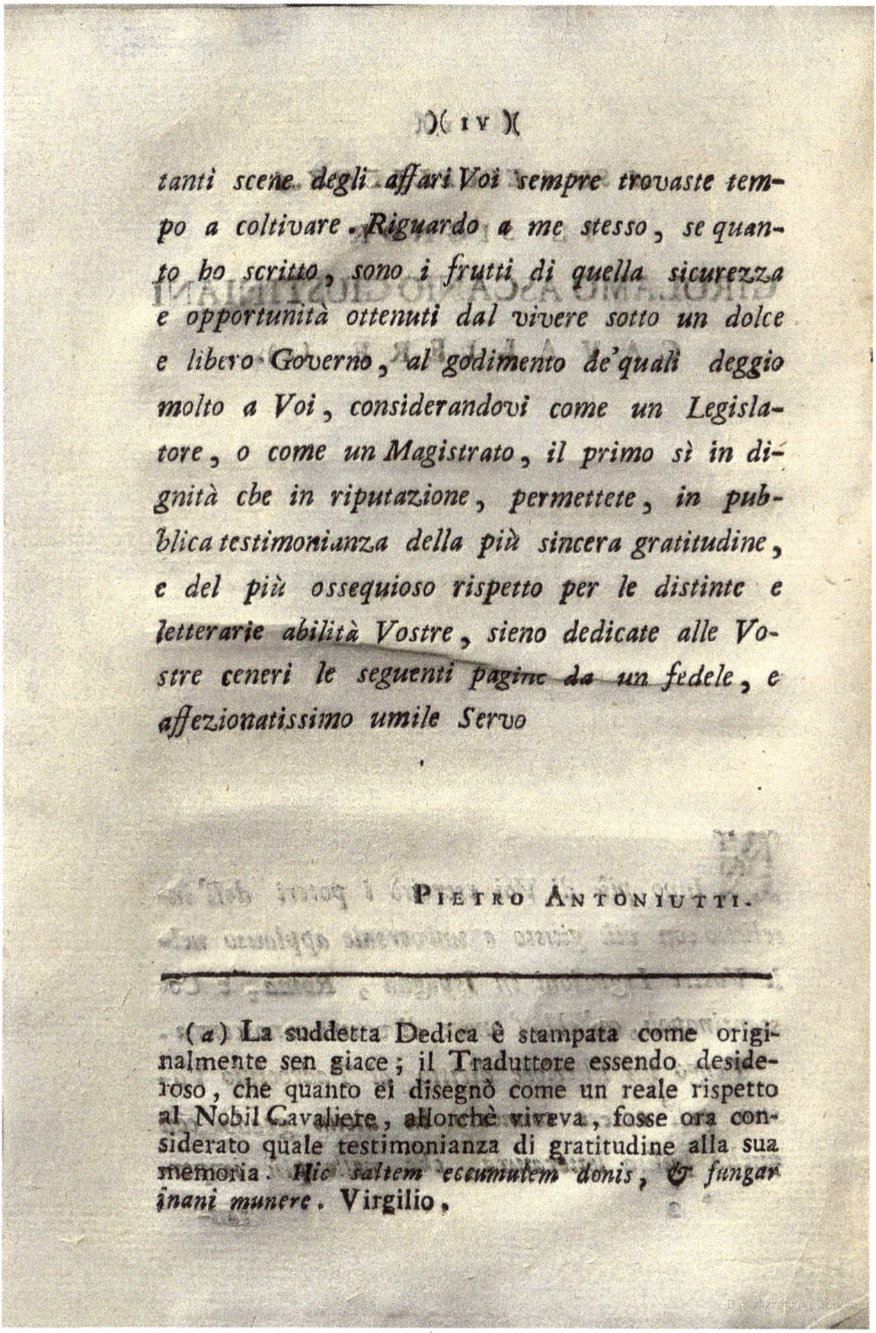

)(IV)(

tanti scene degli affari Voi sempre trovaste tempo a coltivare. Riguardo a me stesso, se quanto ho scritto, sono i frutti di quella sicurezza e opportunità ottenuti dal vivere sotto un dolce e libero Governo, al godimento de' quali deggio molto a Voi, considerandovi come un Legislatore, o come un Magistrato, il primo sì in dignità che in riputazione, permettete, in pubblica testimonianza della più sincera gratitudine, e del più ossequioso rispetto per le distinte e letterarie abilità Vostre, sieno dedicate alle Vostre ceneri le seguenti pagine da un fedele, e affezionatissimo umile Servo

PIETRO ANTONIUTTI.

(*a*) La suddetta Dedica è stampata come originalmente sen giace; il Traduttore essendo desideroso, che quanto ei disegnò come un reale rispetto al Nobil Cavaliere, allorchè viveva, fosse ora considerato quale testimonianza di gratitudine alla sua memoria. *Hic saltem accumulem donis, & fungar inani munere*. Virgilio.

FIGURE 3.1B Antoniutti's dedication, in *Tre lettere apologetiche* (Venice, 1791). Private collection

FIGURE 3.2A William Robertson, *Storia di Scozia* (Londra [Venice], 1784). Private collection

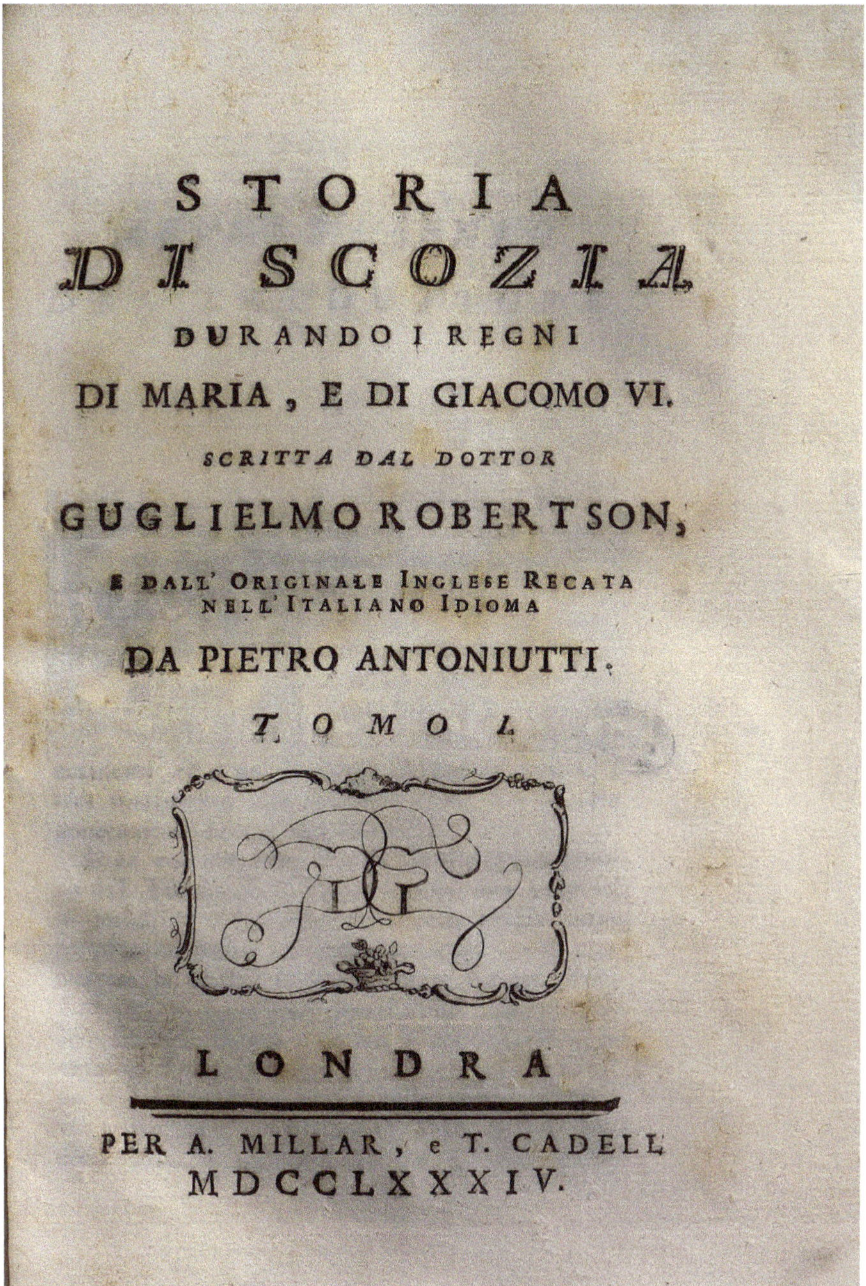

STORIA
DI SCOZIA
DURANDO I REGNI
DI MARIA, E DI GIACOMO VI.
SCRITTA DAL DOTTOR
GUGLIELMO ROBERTSON,
E DALL' ORIGINALE INGLESE RECATA
NELL' ITALIANO IDIOMA
DA PIETRO ANTONIUTTI.
TOMO I.

LONDRA

PER A. MILLAR, e T. CADELL
MDCCLXXXIV.

FIGURE 3.2B William Robertson, *Storia di Scozia* (Londra [Venice], 1784). Private collection

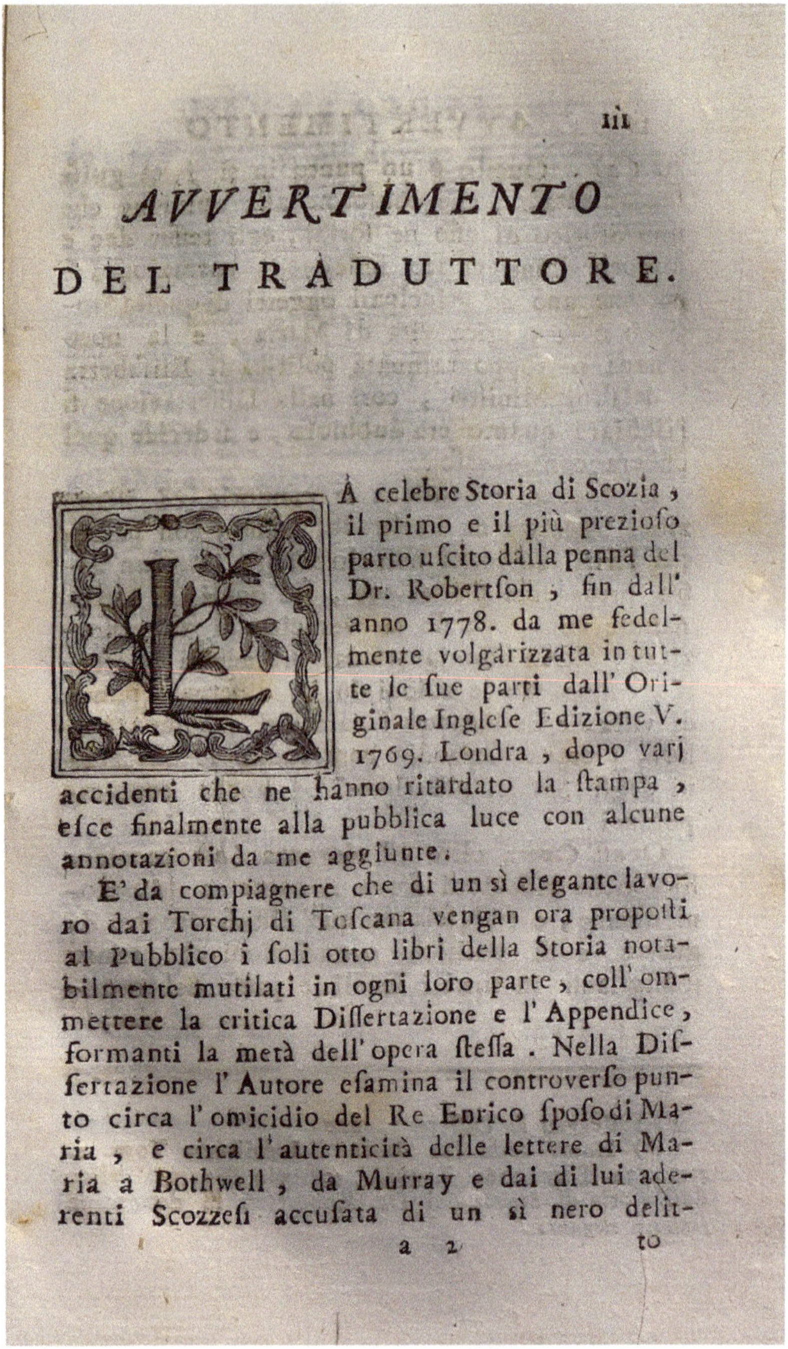

iii

AVVERTIMENTO

DEL TRADUTTORE.

LA celebre Storia di Scozia, il primo e il più prezioſo parto uſcito dalla penna del Dr. Robertſon, ſin dall' anno 1778. da me fedelmente volgarizzata in tutte le ſue parti dall' Originale Ingleſe Edizione V. 1769. Londra, dopo varj accidenti che ne hanno ritardato la ſtampa, eſce finalmente alla pubblica luce con alcune annotazioni da me aggiunte.

E' da compiagnere che di un sì elegante lavoro dai Torchj di Toſcana vengan ora propoſti al Pubblico i ſoli otto libri della Storia notabilmente mutilati in ogni loro parte, coll' ommettere la critica Diſſertazione e l' Appendice, formanti la metà dell' opera ſteſſa. Nella Diſſertazione l' Autore eſamina il controverſo punto circa l' omicidio del Re Enrico ſpoſo di Maria, e circa l' autenticità delle lettere di Maria a Bothwell, da Murray e dai di lui aderenti Scozzeſi accuſata di un sì nero delitto

a 2 to

FIGURE 3.3 Antoniutti's 'Avvertimento del traduttore', in Robertson, *Storia di Scozia* (Londra [Venice], 1784). Private collection

Remaining faithful to the original was, then, a basic value for Antoniutti, and he regularly pointed out that he considered it profoundly wrong to alter an author's thought and mislead the readers as a consequence. For him, it was much better to use footnotes instead; that was the perfect place to engage in a debate with the author and reject the theses that one believed to be wrong or potentially dangerous and misleading. Like forewords, which Antoniutti included in all his publications, footnotes act as paratextual guides for various types of readers. They are carefully thought-out tools meant to help readers orient themselves in the original text. They contain the translator's comments on the main content of the translated works, or supplements aimed at explaining the meaning of specific terms or events described in the text; in addition, they indicate the most important passages to focus on in order to grasp the author's arguments. In addition, Antoniutti even intervened in the very structure of the original publications, adding subtitles to the sections into which the text was divided, once again with the aim of highlighting the narrative's key passages.

The systematic nature of the adaptation operated by the paratext is also evident in the translations of historical works, which make up a considerable part of Antoniutti's production. This category contains, for example, translations of David Hume's *History of England* and William Robertson's *History of Scotland*, published by Giuseppe Parolari and Giovanni Gatti, respectively; these were two potentially dangerous works for a Catholic reader, given the criticism that the two Scottish writers directed at the Church of Rome.

Although both works repay analysis, in the following section I shall concentrate on Antoniutti's translation of Robertson's first historical work.

Antoniutti's Translation of William Robertson's *History of Scotland*

Despite his frequent contacts with British diplomats and travellers, it would be an exaggeration to claim that Antoniutti had a clear and precise understanding of the peculiarities of the Scottish as opposed to the English context. But if we examine closely his translations it is nevertheless evident that he had a certain awareness both of the role that early eighteenth-century Scotland had played in reshaping the power balance within the British Isles and of the repercussions of this situation for European politics. This last issue in particular was of paramount importance in the arguments that Robertson developed in the *History of Scotland*.[31] For the translator, this work was undoubtedly the most

31 William Robertson, *The History of Scotland during the Reigns of Queen Mary and King*

important of the Scottish author's entire production and, in the 'Avvertimento al lettore' that he wrote for his Italian version of the *History*, he drew attention to the fact that

> Gli affari di Scozia sono talmente frammisti a que' delle altre nazioni, e la sua situazione nel politico stato d'Europa è tanto importante, non che l'influenza sì visibile, che l'istoria di que' tempi diviene oggetto di attenzione agli stranieri.[32]
>
> (the affairs of Scotland are so intertwined with those of other nations, its position in the political state of Europe is so important and its influence so visible that the history of those times has become the object of attention for foreigners.)

The desire to publish this translation had been so strong that he had overcome the 'vari accidenti' ('various accidents') that had delayed its printing for almost six years.[33] Although the translation was completed in 1778, as he recalled in the 'Avvertimento', his edition only came off the presses of the Venetian printer Giovanni Gatti in 1784, after being authorised by the Riformatori dello Studio of Padua (the censors of the Republic of Venice), with a false address (London).[34] As we can read in the documents preserved in the Venice State Archive,

---

James VI, till His Accession to the Crown of England, with a Review of the Scottish History previous to that Period, and an Appendix containing Original Papers ... (London: A. Millar, 1759), vol. 2, 315. On Robertson's first work and historical writing see Karen O'Brien, *Narratives of Enlightenment. Cosmopolitan History from Voltaire to Gibbon* (Cambridge: Cambridge University Press, 1997), esp. 129–66; Colin Kidd, 'The Ideological Significance of Robertson's *History of Scotland*', in *William Robertson and the Expansion of Empire*, ed. Stewart J. Brown (Cambridge: Cambridge University Press, 1997), 122–144; Jeffrey R. Smitten, 'Robertson's *History of Scotland*: Narrative Structure and the Sense of Reality', *Clio* 11 (1981): 29–47.

32 'Avvertimento del traduttore', in William Robertson, *Storia di Scozia durando i Regni di Maria e di Giacomo VI, scritta dal Dottor Guglielmo Robertson, e dall'originale Inglese recata nell'idioma italiano da Pietro Antoniutti* (Londra [Venice]: A. Millar-T. Cadell [Giovanni Gatti], 1784), vol. 1, vi.

33 *Storia*, iii.

34 See Gianfranco Tarabuzzi, 'Le traduzioni italiane settecentesche delle opere di William Robertson', *Rivista storica italiana* 30 (1979): 486–509 (489–490), and *False date. Repertorio delle licenze di stampa veneziane con falso luogo di edizione (1740–1797)*, ed. Paola Bravetti, Orfea Granzotto (Firenze: Firenze University Press, 2008), esp. 237. The papers concerning the printing license kept in the *Archivio di Stato di Venezia* (*Dipartimento di Censura*, B 17 f. 43) are interesting and useful for analysing Antoniutti's translation projects.

the censor (Gasparo Gozzi) had initially authorised the printing with a false address on 22 March 1781, but on 29 May of the following year, Giovanni Gatti abandoned the publication. A few days later, the same printing licence was granted to Eugenio Gallici. However, in November 1782, the printer sent another letter of renunciation to the Riformatori dello Studio of Padua. The issue was successfully resolved in August 1783, when the printing licence was again, and permanently, granted to Gatti.

It is possible that Antoniutti's desire to have his work printed – at all costs, one might say – also came from his desire to do justice to the Scottish historian, providing the Italian readers with a vernacular version that was finally complete and respectful of Robertson's historiographical method. In fact, in the very first lines of the 'Avvertimento', Antoniutti expressed severe criticism of one of the previous Italian translations of the *History of Scotland*, published in Siena between 1778 and 1779; he accused it of having presented the eight books comprising the *History* in 'una forma orribilmente mutilata' ('a horribly mutilated form'), deprived of two essential parts: the *Dissertation* and the rich apparatus of 'scritti autentici' ('authentic writings') not found in any previous work on the history of Scotland.[35] The Dissertation at the end of the second volume of the original edition was of particular importance in shedding light on some dramatic episodes in Scottish history (mainly the murder of Henry Stuart, Lord Darnley, Mary Stuart's second husband), while the memoirs and letters served to substantiate the interpretations put forward by the historian, who challenged theses rooted in tradition and in the opinion of Scottish authors.[36] From Antoniutti's point of view, omitting these sections meant compromising the understanding of the *History*'s underlying project. As he recalled at the end of his lengthy opening remarks, Robertson had perfectly interpreted the historian's task of observing and reconstructing the delicate 'mixtures' that distinguished 'the true historical from the realistic and the realistic from the fabulous' and determined 'the different degrees

35 *Storia*, IV.

36 Robertson had decided to publish a series of sources in appendices to the individual volumes of his *History of Scotland*, thus testifying to the rigour and accuracy of his historical reconstructions. In particular, he added an appendix containing *a Critical Dissertation concerning the Murder of the King Henry, and the Genuiness of the Queen's Letter to Bothwell, as well as a selection of the documents used in the drafting of the paragraphs devoted to one of the most delicate issues in modern Scottish history, Lord Darnley's assassination.* As he stated, the main aim of these sections was 'to assist [the reader] in forming some judgement concerning the facts in dispute, by stating the proofs produced on each side' (Robertson, *History of Scotland*, vol. 2, 315).

of strength of the accounts and authority of the witnesses' by making these documents available in the appendix.[37]

In his foreword, the translator also offered a comprehensive overview of the innovative peculiarities of the work he had chosen to translate, starting with the elegant style and ending with the author's ability not to limit himself to a schematic, factual reconstruction of events; instead, he presented more useful philosophical reflections on the dynamics of human action and the logic behind the relationships between different historical events. Closely analysing the content and the structure of the work, the translator dwelled on two aspects in particular: the relevance of a discussion on the origins, progress and decline of the feudal system, on the one hand, and the importance given by William Robertson to the investigation of the protagonists' characters on the other. In the latter case, Antoniutti affirmed his right to insert some entirely personal observations, for example, on the actions of Queen Elizabeth I, who was, in his opinion, too 'rigida e orgogliosa' ('rigid and proud').[38] But he paid most attention to the question of religion.[39] As we can see from the notes and commentaries inserted in his translations of Robertson, Hume and other British philosophers and historians, Antoniutti firmly believed in the purity of the original essence of the Catholic religion and pointed out the deviations and excesses committed throughout the secular history of the Church, distinguishing between just doctrines and the flawed actions of the clergy. In his notes, the translator offered his own analysis of the characteristics of the Anglican Church and the Church of Scotland, in which he also highlighted the dangerous consequences of too violent hostility to the Catholic clergy. Antoniutti considered it indispensable to promote 'sound principles of religion and virtue', which would help a civil and virtuous development of society. Thus he credited Robertson's historical work for distinguishing clearly on numerous occasions such principles from the opposed, unruly conduct of individual representatives of ecclesiastical institutions.[40] Nonetheless, he stressed how the *History of Scot-*

37 'I miscugli [che distinguono] il vero storico dal verosimile e il verosimile dal favoloso, determinando i diversi gradi di forza delle testimonianze e di autorità de' testimoni', *Storia*, xii–xiii.

38 *Storia*, vii. On the role of biographical portraits in Robertson's work see Neil Hargraves, 'National History and Philosophical History Character and Narrative in William Robertson's *History of Scotland*', *History of European Ideas* 26, no. 1 (2000): 20–33.

39 Alexander Du Toit, 'A Species of False Religion: William Robertson, Catholic Relief, and the Myth of Moderate Tolerance', *Innes Review* 52 (2001): 167–188. On the 'religious question', see also Mary Fearnley-Sander, 'Philosophical History and The Scottish Reformation: William Robertson and the Knoxian Tradition', *The Historical Journal*, 33 (1990): 323–338.

40 See 'Avvertimento del traduttore', in *Storia*, esp. xii–xiii.

land was not without serious flaws, including expressions that 'sembrassero ingiuriose alla Fede' ('seemed insulting to the Faith'). In some cases they were inevitable, given the tragic nature of the periods dealt with in the narrative; at other times, they were a mere reflection of the positions of 'heterodox Englishmen' who, still in the eighteenth century, defended theses 'a thousand times refuted by our Catholic writers'.[41]

Antoniutti intervened in these passages in particular with a series of 'warning' notes placed in various places in the text, though without ever rewriting or censoring it. This strategy was similar to that adopted by the French translator of the *History*, Nicolas de La Chapelle,[42] and was much more respectful of Robertson's aims; in a letter to the translator Jean-Baptiste Suard, the Scottish historian had expressly asked that his texts not be altered, but rather supplemented with any necessary notes to adapt them to a Catholic country.[43] The first of these notes – and the only one in Book I – contained a critique of Henry VIII's decision to oppose the Papacy, and developed into a harsh attack on Celsi, Giuliani, Sozzini, Bayle and all those who sought to 'build the school of error and perpetuate the lineage of the ungodly'.[44]

The most conspicuous series of interventions obviously came in Book II, which included sixteen detailed footnotes. The most elaborate ones concerned the activities of John Knox, Martin Luther and the other 'riformatori novatori' (innovating reformers) who printed books filled with 'espressioni assurde' (absurd expressions), errors and insults, presenting them as absolute truths.[45] In one of those notes we read that

> La vera o pura Cattolica Religione ... cresce più evidente e poderosa da ogni attacco che facciano sopra di essa; e converte l'amarezza e il veleno de' suoi avversari in nutrimento a se stessa, e in universale rimedio alle pene e ai malori di un misero mondo digenerato.[46]

---

41 'Inglesi eterodossi che ancora nel XVIII secolo sostengono tesi ben mille volte confutate dà nostri cattolici scrittori', *Storia*, xiii–xiv.

42 William Robertson, *Histoire d'Ecosse, sous les règnes de Marie Stuart, et de Jacques VI ... traduite de l'anglois* (London [Paris]: n.n., 1764).

43 On this point see John Renwick, 'The Reception of William Robertson's Historical Writings', in *William Robertson and the Expansion of Empire*, 145–163.

44 'Tutti coloro i quali cercavano di erigere la scuola dell'errore e perpetuare la stirpe degli empi', in *Storia*, 1784, 75.

45 *Storia*, p. 138. Antoniutti dedicated four extensive notes to the refutation of the ideas of the Protestant reformers (*Storia*, 150–156).

46 *Storia*, 147.

(The true or pure Catholic Religion grows more evident and powerful with every attack they make upon it; and it converts the bitterness and poison of its adversaries into nourishment for itself, and into a universal remedy for the pains and sorrows of a miserable and degenerate world.)

This 'miserable and degenerate world' contested both ecclesiastical celibacy – which Antoniutti considered a great virtue, even if it was not a rule established in the sacred texts[47] – and excommunication, whose uses and characteristics were duly reconstructed in an extensive note.[48]

The other arguments that were refuted concerned the 'vices' of Catholic priests – which were completely unfounded and instead more applicable to the English clergy than to the Catholic[49] – or the 'assurdi, malfondati e speculativi errori' ('absurd, ill-founded and speculative errors') that characterised Catholic doctrine according to the Scottish historian.[50] An interesting annotation can be found in Book IV, where the murder of the Italian Davide Rizio is narrated,[51] eliciting severe criticism of the historian; Robertson was accused of 'assurde esagerazioni contro il Cattolicesimo' ('absurd exaggerations against Catholicism') and of failing to remain fair and impartial, as promised in his Preface.[52]

This rich insertion of comments and observations did not correspond to any particular rewriting of the text, which was translated into Italian as faithfully as possible. The work's overall structure was respected in every detail, from the paragraph headings to the footnotes, which contained all the original bibliographical references. Even the final index resembled the English one; it presented a list of all the people and places cited in the volumes, accompanied by brief explanations of the aspects dealt with for each entry.

Thus the fidelity that, according to Antoniutti, was a salient feature of his translation was reflected in his complete respect for the source edition and the author's theses, which were only countered in the notes – his 'spaces of dialogue' with the reader. He chose to remain faithful to the original text even

47 *Storia*, 158–159.

48 *Storia*, 161–162.

49 *Storia*, 163–165.

50 *Storia*, 208–209. In Book IV, for example, to counter Robertson's opinions on the political role of the Roman clergy, the translator developed a brief reflection on the Catholic Church's propensity for forgiveness, benevolence and piety (*Storia*, vol. 2, 1784), 8.

51 David Rizio, a poet and composer of Piedmontese origin, became part of Mary Stuart's court. He was assassinated in April 1566 by Lord Darnley, who feared his relationship with the Queen and the influence he exerted on her decisions.

52 *Storia*, vol. 2, 22.

in this case, rejecting the solutions adopted by other printers. He preferred to criticise Robertson's ideas point by point in the footnotes, relying on the fact that, thanks to the guidelines provided in his notes or forewords, the readers themselves could become aware of what might be false and incorrect in certain theses or arguments. Further still, he structured the translation in such a way that he directed the readers to the strengths of the *History*, leaving aside the parts that focused on theological or ecclesiastical issues. For him, it was more important to follow Robertson's line of reasoning regarding the dynamics inherent in the birth of modern states than to get lost in the pages describing the figure of John Knox, where the value of the sacraments was questioned or the importance of the contributions of heretical authors – such as the Sozzinis or Pierre Bayle – was acknowledged by the Scottish historian.

This distinctive feature of his translation style was recognised in numerous reviews, such as the one that featured in the *Giornale italiano letterario* in 1812, in which his commentaries and notes were defined as 'antidoti' ('antidotes') to counter the 'veleno' ('poison') of philosophers or other Protestant authors.[53] Antoniutti's translation of Robertson's *History* was also welcomed in the article published in the Venetian journal *Progressi dello spirito umano* mentioned above. According to the anonymous reviewer, the 'excellent history' written by the Scottish historian, which had literally won over English readers, would also arouse the curiosity of Italian scholars; they had always been interested in British affairs, but their desire for knowledge was hindered by a 'doppia difficoltà' ('twofold difficulty'), due to both the complexity of the events themselves and the widespread prejudice that Protestant historians were not objective when dealing with issues involving the Catholic Church and religion. The translation of the *History* was of great help in overcoming both obstacles, as it allowed the circulation of a work where facts were recounted with exemplary seriousness and clarity, and where all the letters, reports, memoirs necessary for an accurate description of them had been used and brought to light. The Scottish historian had spared no effort in researching every possible document and testimony, reviewing the material held in libraries and private collections, thereby ensuring his work reached a new level of excellence. Such material had been generously made available to the public and, according to the journalist, Antoniutti had been very careful to translate every single page and document. The learned

53 *Giornale italiano letterario*, 192. See also the review published in *Progressi dello spirito umano nelle scienze, e nelle arti*, 18 February 1784, 195–197. Both of them were anonymous but signed with the initial A. In all probability, the reviewer was the priest and scholar Tommaso Antonio Contin; see Del Negro, *Il mito americano*, 152, and Marino Berengo, *Giornali veneziani nel Settecento* (Milan: Feltrinelli, 1962), xxiv–xxvi.

translator had decided not to change the structure of the text, translating it in its entirety while, at the same time, adding precise annotations and comments to his edition in order to guide less experienced readers who may be more susceptible to Robertson's 'idee alterate' ('distorted ideas') on religion. Also significant, in my opinion, was the journalist's concluding statement, arguing that prohibiting the translation of works containing ideas that were not good for a Catholic reader was never a good choice for the advancement of knowledge; it was not difficult to guide and protect readers if a translation model such as the one proposed for the *History of Scotland* was used.

As mentioned above, a very similar strategy of peritextual modification was also adopted in the case of Hume's even more polemical *History of England*, in which the translation was enriched with dozens of detailed notes rejecting the theses of the Scottish historian and philosopher and suggesting more edifying readings for a Catholic audience.[54] It should however be noted that this strategy caused Antoniutti many problems. His firm rejection of any form of censorship frequently led him to clash with the Riformatori dello Studio of Padua, the censors of the Republic of Venice. Indeed, as some nineteenth-century biographers such as Girolamo Dandolo or Luigi Narducci recounted, in some cases the Riformatori denied permission to the printers contacted by Antoniutti to publish his work, with the result that his projects remained in the form of manuscripts. In other cases, on the contrary, they granted permission provided that several additional notes be included, written directly by them. They often took up to six years to grant this permission, forcing Antoniutti to write petitions and letters reminding them of his requests for permission to publish. In still other cases, the Riformatori intervened during the project, blocking the printing of an already authorised translation. This is what happened to Antoniutti's version of the *History of England*; he managed to publish only four out of the six planned volumes, as the Riformatori dello Studio of Padua judged that the notes he was adding to counter Hume's dangerous claims were insufficient.

From a more general point of view, we can highlight that the rich and multifaceted apparatus of notes and commentaries added by Antoniutti to his translations could be justified not only by the need to warn readers about

54 David Hume, *Istoria dell'Inghilterra di David Hume dalla invasione di Giulio Cesare sino alla rivoluzione 1688. Volgarizzata dall'abate Pietro Antoniutti* (Venice: Parolari, 1819–1820). Antoniutti added to his edition the translation of 'My Own Life', the autobiography of Hume with the addition of a letter written by Adam Smith and dated November 9, 1776. On the Italian reception of Hume's works cf. Emilio Mazza, 'Translations of Hume Works in Italy', in *The Reception of David Hume in Europe*, ed. Paul Jones (New York: Thoemmes Continuum, 2005) and Maria Luisa Baldi, *David Hume nel Settecento italiano. Filosofia ed economia* (Firenze: La Nuova Italia, 1983).

passages that were dangerous in moral or religious terms. They could also be used to improve on the original by clarifying terms and concepts used by Robertson, Hume and other British authors; the additional explanations placed the work more firmly in the target context, to give readers a better understanding of the content, and the further bibliographical references often concerned texts or other translations by Antoniutti himself. This strategy was adopted, for example, in Antoniutti's version of *Travels into Poland, Russia, Sweden and Denmark* by the English historian William Coxe, published in London in 1784: the chapters containing information on Constantinople and the relationship between Europe and the Ottoman empire were enriched with comments and information on Ottoman history and culture.[55]

Final Translation Projects

The career of Pietro Antoniutti continued even after the signing of the Treaty of Campoformio, which brought an end to the Republic of Venice, and during the two Austrian occupations that followed.[56] Despite the changed political and institutional situation, he persisted in his attempts to stimulate the thinking of the Venetian aristocrats and readers, even if, as we might expect, he encountered various difficulties and had to deal with Austrian censorship. This phase is therefore marked by a greater diversification of his production: on the one hand, it no longer made sense to choose many texts about the decline of the republics, given that the Venetian government was by then neither 'gentle' nor 'free'; on the other hand, the need arose to find points of reference that might help readers understand what was happening in the European continent on the political level. More generally, Antoniutti felt the time had come to examine how modern societies evolved; he therefore dedicated himself to the translation of Adam Ferguson's *Essay on the History of Civil Society* (1807)[57] and the philosophical and historical works of Henry Bolingbroke, James Harris, David

55 William Coxe, *Viaggi in Europa del sig. Guglielmo Coxe inglese. Compendio della storia generale de' viaggj continuazione dell'opera di M. de La Harpe .... Adorna di carte geografiche e figure, arricchita d'annotazioni ...* (Venice: Tosi, 1790–1792).

56 On the institutional and political situation of Venice at the beginning of the nineteenth century see Michele Gottardi, *L'Austria a Venezia. Società e istituzioni nella prima dominazione austriaca, 1798–1806* (Milan: FrancoAngeli, 1993).

57 Adam Ferguson, *Saggio circa la storia di civile società di Adamo Ferguson* LL.D. *professore di filosofia morale nella Università di Edinburgo, volgarizzamento di Pietro Antoniutti* (Venice: Santini, 1807).

Hartley, and George Lyttelton, among others.[58] Particularly interesting was his choice of publishing a miscellany including works of the best-known English writers, *Sinopsi istorico-cronologica de' più celebri autori inglesi*,[59] to which he decided to add an additional work, *Calcolo sopra la verità della storia* written by the philosopher and scientist Giovanni Maria Ortes. In so doing, Antoniutti wanted to stress the importance of the contribution made by Italian and Venetian men of letters to the development of the eighteenth-century debate on historiography. Even though the Republic had lost its political independence, he was demonstrating that Venice continued to play a prominent role within European culture.

Furthermore, at this stage of his career, he developed a new interest in more purely linguistic and literary works. Picking up on suggestions from the Milanese debates on the usefulness of translations, which would reach a turning-point in 1816 after the publication in the *Biblioteca italiana* of the celebrated article 'Sulla maniera e la utilità delle traduzioni' by Baroness Anne-Louise-Germaine de Staël-Holstein,[60] Antoniutti tried his hand at the translation of works concerning the evolution of national languages and literatures. One example is the publication of the *Saggio sopra gli scritti e 'l genio di Shakespeare paragonato a Greci e Francesi drammatici*,[61] translated from a work by Lady Elizabeth Montagu, the well-known co-founder of the Blue Stockings Society.[62] This work was accompanied by a dedication to Lady Foscarina Giovannelli (wife of Leonardo Manin, nephew of the last Doge of Venice),[63] as well as by a long foreword, where Antoniutti presented in detail his ideas on the social and

58 See, for example; Henry Bolingbroke, *Riflessioni sopra la storia d'Inghilterra* … (Venice: Andrea Santini e figlio, 1814); James Harris, *Filologiche ricerche del cavaliere Giacomo Harris* … (Venice: Santini, 1809); David Hartley, *Osservazioni sopra l'uomo, sua struttura, suoi doveri, e sue speranze di David Hartley A.M.* … (Venice: Santini, 1807); George Lyttleton, *Storia d'Inghilterra dai più rimoti tempi fino all'anno 1784* … (Venice: Santini, 1807).

59 *Sinopsi istorico-cronologica de' più celebri autori inglesi dalla invasione di Giulio Cesare fino ai giorni nostri* …, (Venice: Santini, 1806). See Castagnino, 'La plume du traducteur', 63–64.

60 'Sulla maniera e la utilità delle traduzioni', *Biblioteca italiana, ossia giornale di letteratura scienze ed arti compilato da una società di letterati* 1, (January 1816), 9–18. See Mirella Agorni, *Translating Italy for the Nineteenth Century. Translators and an Imagined Nation in the Early Romantic Period 1816–1830s* (Bern: Peter Lang, 2021). An overview of the Italian debate on Mme de Staël's article is provided by Riccardo Massano, 'Finalità e caratteri del tradurre nel pensiero dei primi romantici italiani', *Atti della Accademia delle Scienze di Torino* 94 (1959/1960), 347–403.

61 Montagu, *Saggio sopra gli scritti e 'l genio di Shakespeare.*

62 See Elizabeth Eger, *Bluestockings: Women of Reason from Enlightenment to Romanticism* (Basingstoke: Palgrave Macmillan, 2010).

63 The letter of dedication offered Antoniutti an opportunity to celebrate the role played by the Manin family in the glorious history of Venice.

cultural importance of translations, which included some issues that are still at the heart of debates in Translation Studies today. In particular, the main focus of Antoniutti's argument was on the function and role of translators, who had an arduous task; instead of being a mere linguistic adaptation, their work was aimed 'ad ampliare delle pratiche cognizioni' ('at broadening practical knowledge') in every field, from literature to botany, from 'le scienze più raffinate' ('the most sophisticated sciences') to history.[64] In his opinion, translation was a high-level cultural process, a channel for introducing new authors and enriching the language of the target culture, and, at the same time, a tool for educating new readers from different social and cultural backgrounds.[65]

Antoniutti's career probably ended in 1820. We have no further information about his translation activities after that date. In the biographies published in newspapers and literary journals, a portrait emerges of a man of letters who was very careful in his choice of texts to translate, even if he was not always precise in his lexical and grammatical choices. As argued by Girolamo Dandolo, he was not a genius, but a 'uomo di valore' ('man of value'), able to offer readers translations of all the most interesting and useful works on the market.[66]

The picture that emerges from the available biographical sources and the examples of specific translation strategies described here is that of a scholar and intellectual ahead of his time, who developed a well-defined cultural and, to some extent, political project. A project based exclusively on the adaptation of European works, rather than on the production of original writings. Less interested than other translators of his time in discussing theoretical issues, Antoniutti focuses all his efforts on a series of specific translation strategies, ranging from very minor textual intervention to the potentially radical transformation produced by the paratext. This enabled him to promote the circulation of versions that were adapted to his project of informing and educating the Venetian reading public. As his biographers observed and as he himself pointed out in his forewords and commentaries, the exceptional or 'heroic' nature of Antoniutti's work consisted in his considering translations as cultural tools with a strong 'civic' and 'social' value.

---

64 Montagu, *Saggio sopra gli scritti e 'l genio di Shakespeare*, vii.

65 Montagu, *Saggio sopra gli scritti e 'l genio di Shakespeare*, 9. In his 'Avvertimento', Antoniutti discusses briefly the contemporary debates on translation.

66 Dandolo, *La caduta della Repubblica di Venezia*, 213–214. See also *Giornale dell'italiana letteratura* 24 (1810): 175–180, review of *Avvisi del Lord Chesterfield a suo figlio sopra di uomini e di costumi per un nuovo sistema di educazione, volgarizzamento di Pietro Antoniutti* (Venice: Santini, 1810), and the review of the *Saggio metafisico, ossia ipotesi circa la formazione e generazione degli enti spirituali e materiali, volgarizzamento di Pietro Antoniutti* (Venice: Santini, 1811) published in the *Giornale italiano letterario* 33 (1812): 191–192.

CHAPTER 4

Joseph Morgan between England and Algiers

Ann Thomson

> [...] one almost daily meets with People, of both Sexes, ruder, more brutal, less polished, and superlatively more abandondly impious, in this noble City, this very Metropolis of ours, than, I positively aver, I ever met with in Barbary, during my near twenty Years Intercourse in that savage Country; even among the wildest of the Arabs and Africans.[1]

The person who wrote these lines in London, a certain Joseph Morgan, is an obscure figure about whom we know very little, whose first name is uncertain[2] and the details of whose life, including his dates of birth and death, are unknown. The brief and rather inaccurate article in the *Oxford Dictionary of National Biography* simply lists his works and some of his translations, based on the British Library catalogue. The few passing references to him by earlier historians see him as an unscrupulous translator who used the works of others, and he was accused of being an 'indefatigable plagiarist'.[3] However, as I hope to show, Morgan's life and publications, including in particular his translations, the result of a long sojourn on the southern shore of the Mediterranean (or rather Africa, as he always termed it), repay a more detailed study. He exemplifies not only the effect of mobility on translation but also the role played by Moslem countries and Islam in European intellectual culture in the eighteenth century.[4]

Almost all the information that we have about Morgan comes from his own writings. He was presumably born in the late seventeenth century; he claims

1 Joseph Morgan, *A Complete History of Algiers*, vol. I (London: printed for the author, 1728), viii.

2 His title pages indicate 'J. Morgan, Gent.'. His letter to Prideaux reproduced in *Mahometism Fully Explained* (London: E. Curll, W. Mears, and T. Payne, 1723), ix, is signed Jos. Morgan, but the *Gentleman's Magazine*, November 1732, 1081, refers to him as John Morgan.

3 Robert Lambert Playfair, *Bibliography of Algeria, from the Expedition of Charles V in 1541 to 1887* (London: John Murray, 1888), 150; see also Magali Morsy, *La Relation de Thomas Pellow. Une lecture du Maroc au 18e siècle* (Paris: Recherches sur les civilisations, 1983), 14.

4 Some of this information was discussed in a different form in my article 'Joseph Morgan et le monde islamique', *Dix-huitième Siècle* 27 (1995): 351–363; see also Nabil Matar, 'Joseph Morgan'

 | DOI:10.1163/9789004731806_006

that after having spent twenty years in Algiers, where he was 'chancellor' to the British Consul Robert Cole (who died in 1713),[5] he left in 1719 for Tunis before returning to England in 1720.[6] The earliest information we have about him is that he was present at the siege of Oran in 1707.[7] On his return to England, he published several books on North Africa, in particular his relatively successful *Compleat History of Algiers*,[8] but mainly seems to have earned a living as translator; amongst other things he translated not only works on North Africa and Spain but also an early seventeenth-century Morisco manuscript on Islam, and several writings by the Italian freethinker Alberto Radicati di Passerano. The publication of the latter's *Philosophical Dissertation on Death* led in 1732 to the arrest of author, translator and bookseller.[9] Morgan also published in 1732 a collection of seventeenth-century pamphlets in the first (and only) volume of *Phœnix Britannicus*, dedicated to the Duke of Richmond, Grand Master of the Freemasons, in which Morgan declared his membership of the Masons and requested the support of his fellow Masons.[10] He launched in 1730 a political periodical called *The New Political State of Great Britain. Including the public affairs of foreign courts*, only the first volume of which seems to have appeared. His last publication was *The Lives and Memorable Actions of many Illustrious Persons of the Eastern Nations* (1739), a work planned by the Arabist and famous translator of the Quran George Sale and left unfinished on the latter's death. After this date nothing more is heard of Morgan. An English version of Laugier de Tassy's 1725 French work on Algiers, very probably translated by Morgan, appeared in London in 1750 under the title *A Compleat History of the Piratical*

in *Christian-Muslim Relations, A Bibliographical History* ed. David Thomas and John Chesworth, vol. 12 (Leiden: Brill, 2020).

5 J. Morgan, *A Complete History of Algiers* (London: A. Bettesworth and C. Hitch, 1731), Preface, v.

6 *Several Voyages to Barbary* (London: Olive Payne, Joseph Duke and Samuel Baker, 1736), 113; *Complete History of Algiers*, 315.

7 *A Compleat History of the Present Seat of War in Africa* (London: W. Mears and J. Stone, 1732), 22.

8 The first edition was published by subscription: *A Complete History of Algiers. To which is prefixed, an epitome of the general history of Barbary, from the earliest times* (London: printed for the author, 1728–1729); a second edition was published by two booksellers in 1731 (see note 5).

9 See Franco Venturi, *Saggi sull'Europa illuminista I: Alberto Radicati di Passerano* (Turin: Einaudi, 1954), 209 ff. He was apparently too poor to pay the bail money himself.

10 *Phœnix Britannicus, being a miscellaneous collection of scarce and curious Tracts* (London: T. Edlin and J. Wilford, 1732); see H. Sirr, 'J. Morgan and his "Phœnix Britannicus", with notes about his other works', in *Trans. Quatuor Coronati Lodge* 19 (1906): 127.

States of Barbary ..., by a gentleman who resided there many years in a public character. As it nowhere indicates the name of the author (or indeed of Morgan), he has been accused of plagiarism;[11] however, there is no indication that he was responsible for this publication, which does not correspond to his usual practice of indicating his name as either author or translator.[12] Thus Morgan's known intellectual activity dates from 1724 to 1739, fifteen years during which, despite his best efforts, he seems to have remained a marginal figure and not to have achieved the intellectual recognition or the patronage he was apparently seeking. He has since been almost totally forgotten. These meagre details are very intriguing and in themselves invite further investigation.

We have no information at all concerning his early life or the reasons which led him to Algiers, or concerning his relations with Consul Cole, who apparently arrived there after having been taken captive by corsairs. Nor do we know what Morgan did there between Cole's death and his own departure six years later. But on his arrival in England in 1720 he immediately set out to make a name for himself in literary circles and to seek patronage. This is clear from the work he published in two volumes, in 1723 and 1725: *Mahometism Fully Explained*, which was the translation of a manuscript poem by Mohamed Rabadan de Aragón (1580?–?), *Discurso de la luz*, which presents a biography of the Prophet, based on Arab authorities but directed towards the exiled Morisco community. The first volume is dedicated to Lord Harley; Morgan explains that the work is the translation of a manuscript that he had given to Harley for his collection.[13] The second volume, *Mahometism Explained*, is dedicated to the influential Dr Richard Mead, a leading physician and patron of scientists and artists, who had a huge collection of books and manuscripts and shortly afterwards became physician to King George II.[14] This second volume contains an account of the sufferings of the Moriscoes (the Muslims expelled from Spain), in the course of which Morgan provides extracts from another manuscript that he had borrowed while in

11 See note 3.

12 See A. Thomson, 'Joseph Morgan et le monde islamique'; unlike N. Matar, I therefore do not include it in the list of Morgan's publications.

13 For information about Rabadan and his work, see Nabil Matar, 'Jos.[eph] Morgan and Mohamed Rabadan: the First Muslim Biography of the Prophet Muhammad in English', *Journal of Islamic Studies* 30 (2019): 151–175.

14 Anita Guerrini, 'Richard Mead', *Oxford Dictionary of National Biography* (Oxford, New York: Oxford University Press, 2004, 2022). https://www.oxforddnb.com/view/10.1093/ref:odnb/9780198614128.001.0001/odnb-9780198614128-e-18467.

North Africa; it belonged to a Moor whose grandfather, the author, had been expelled from Granada in 1610.

Mahometism [*Fully*] *Explained*

The translation of Mohamed Rabadan's work was Morgan's first publication. He explained that it was the translation of a rare manuscript written in Spanish in Arabic characters that he had bought in September 1715 at Tessatore, a small town west of Tunis, from a certain Hamoda Bussisa Tibib, a barber surgeon who, he says, is one of the few who can still read Spanish.[15] He also insists several times on the great prestige enjoyed by its author Mohamed Rabadan among North African Moors and on the popularity of the work in Tunis.[16] Before publication he had sent his translation to the highly respected churchman Dean Humphrey Prideaux, who had written a polemical life of the Prophet called *The True Nature of Imposture Fully Displayed in the Life of Mahomet* (London, 1697). However, Prideaux was too ill to give his opinion and in fact died soon afterwards.[17] Morgan claimed that as the manuscript was written in Spanish in Arabic characters, in a mix of different Spanish dialects with Arabic words spoken by the Moriscos in North Africa, he was one of the few people capable of understanding it, because 'by my long Continuance, and frequent Conversations with the Off-spring of those Exiles in Barbary, I am thoroughly vers'd in their Way of expressing their Sentiments.'[18] This, together with the fact that he was addressing Prideaux, indicates that he wanted to make a name for himself as a specialist, while the claims on the title page of the book that the work contains 'many Surprizing Passages, not to be found in any other Author' obviously reveal commercial aims. In this, however, he was not successful. He refers in the Dedication to Mead in volume II of his translation of Rabadan to criticisms made of his publication of the work, and he did not publish the follow-up which – as he indicated in the 'Advertisement' at the end of the same volume – he intended to publish depending on 'the Reception the Town gives to this'. Apparently the reception was not favourable, possibly due to the hostility he

15 *Mahometism Fully Explained* (London: E. Curll, W. Mears, and T. Payne, 1723), 'The Author's Preface', note, [6].

16 See for example *Mahometism Explained*, vol. II (London: Printed for the Author, 1725), vii–viii and *Mahometism Fully Explained*, I, [5].

17 Morgan reproduces the correspondence at the beginning of *Mahometism Fully Explained*, I, vii–x.

18 Letter to Prideaux, *Mahometism Fully Explained*, I, viii.

mentions from people who were opposed to an account of Islam that was not solely critical or part of Christian polemic.

It is also relevant to note that while the first volume was published by three established booksellers, Morgan had to publish the second volume himself by subscription, which indicates that no bookseller was willing to take it on, although the list of subscribers does include aristocrats, politicians, Church dignitaries and well-known writers. Morgan himself makes very clear, in the Preface to the first edition of his *History of Algiers*, also published by subscription, how demeaning it is to hunt for subscriptions, and even announces that he will publish 'The History of the Progress of the History of Algiers; being a Fragment of, or rather An Essay upon the Comi-Tragical History of Subscription-Hunting'.[19] Unfortunately, I have found no trace of the publication of such an essay, but this Preface does contain some indignant remarks about the attitude and ignorance of the reading public, in the course of which he refers to himself as 'one dwindled to accost, Cap-in-hand, under the scandalous, pitiful and contemptible Character of *Subscription-Hunter*. An abject Vocation! Here especially'.[20] It is perhaps for this reason that he moved on, after his publication of the translation of Rabadan, to publish this history of Algiers in 1728. Despite his castigation of the reading public that we have just seen, this genre appealed more to the tastes of the general reader, as it cashed in on the fascination for accounts of foreign countries, particularly dealing with a country seen as both exotic and barbarous, due to the activity of its corsairs in the Mediterranean. That he was right can be deduced from the fact that while the first edition, in 1728, was published by subscription, there was a second edition in 1731, published by established booksellers.

Morgan's translation of Rabadan's text is accompanied by notes, some commenting on the interpretation of the text and some explanatory, with information taken from recognised specialists like Humphrey Prideaux, as well as Barthélemy d'Herbelot's *Bibliothèque orientale*, or the writings of Simon Ockley (Professor of Arabic at Cambridge University) and Adriaan Reland (Professor of Oriental Languages at Utrecht University).[21] He also provides information acquired from his own interlocutors in North Africa and, to underline the importance of Rabadan's work for the Moriscoes, he adds a note again emphasising his own first-hand knowledge: 'At the Town of Tessatore, in the Kingdom

19 *A Complete History of Algiers*, x.

20 *A Complete History of Algiers*, xvi.

21 For more information on these specialists, see in particular Alexander Bevilacqua, *The Republic of Arabic Letters. Islam and the European Enlightenment* (Cambridge MA: Harvard University Press, 2018).

of Tunis, I heard some of the Inhabitants, of both Sexes, sing, in Concert, whole Chapters out of this Work, to the Sound of Lutes and Guitars'.[22] He prefixed to his translation the articles of faith of the Moslems, explaining in a note that this was the English translation of another manuscript written in Spanish in Arabic characters conserved in a public library in Amsterdam; it had recently been published in French from the Latin version, translated by David Durand, together with Durand's translation of Adriaan Reland's work on the Muslim religion published in 1721.[23] In all of these preliminaries we can see, as in the dedications, Morgan's attempt to situate himself in the circles of erudite and respected scholars of Arabic and the Muslim world. This is also visible in Morgan's reference to another manuscript, which is particularly intriguing. In the second volume of his translation he provides some translated extracts from a manuscript which he claims to have seen when he was in North Africa. It was lent to him by a Spaniard (who, he insists, was a moderate Catholic), to whom it had been lent by a Moor, the grandson of the author, who could not read it. It was written in 1615 by a Morisco expelled from Spain in 1610, called Abdelkrim Ben Aly Perez, and the first part contained an account of the Muslim religion with a list of the Prophet's miracles, while the second part attacked the Catholic church's beliefs and practices. Morgan provides several long extracts, commenting, 'I much regret my not being Master of the Whole, it being well worth publishing'.[24]

In his Dedication to Dr Richard Mead of the second volume of his translation of Rabadan, Morgan refers to the physical manuscript that he had presented to Lord Harley, saying that was therefore 'in the very best of company [...] in one of the most valuable private Libraries in *Europe*; though in its uncouth *Mauritanian* garb, it makes but a very indifferent figure'. He attributes to his own efforts the fact that it had managed to travel as far as England, and had been 'made intelligible in this Island':

> since the idiom in which it is written, is an unaccountable *Jargon*, peculiar to the *Spanish Moors*; and, I may venture to assert, that scarce one *European* in a Million is capable of extracting the real Sense of ten Pages of the whole Manuscript.[25]

---

22 *Mahometism Fully Explained*, v.

23 *La religion des Mahométans, exposée par leurs propres Docteurs ..., tiré du latin de M. Reland ...* (La Haye: I. Vaillant, 1721), LI ff.

24 *Mahometism Explained*, II, 300.

25 *Mahometism Explained*, II, v–vi.

Morgan also reinforces the veracity of his work and his own competence by referring to his personal discussions with North Africans, both when he was in Algiers and with the entourage of various envoys from different Barbary states to London. He names a certain 'Yusouf Hogia', an envoy from the Bey of Tunis to the English court, whom he quotes at length on the comparative status of women in Christian and Muslim countries.[26] Similarly, on the title page of his *History of the Present Seat of War in Africa* (1732), he refers to conversations he had had with Haj Mahammed, the Algerine Envoy to London, and his Secretary Haj Ali. In this latter work he corrects mistakes made by earlier historians of Oran.[27]

Morgan's translation of Rabadan has recently been studied by Nabil Matar, who compares it to the original and both brings out the particularities of Rabadan's account of the life of the Prophet, written for Moriscoes, and discusses Morgan's point of view as well as aspects of his translation.[28] He emphasises the importance of this publication as representing the first Muslim biography of the Prophet published in English. I shall therefore not go into details concerning the precise content of the work or its representation of Muhammad, but would instead like to situate it in the context of Morgan's translation activities. Despite his deference towards Dean Prideaux, whose polemical life of Muhammad was a work of Anglican propaganda against both Islam and free thought at home, Morgan's aim in this translation, as in other works discussed below, was in fact to undermine European prejudices against Muslims. This is clearly spelled out later, in the Preface to his *History of Algiers*, in which he translates the comments by the French consul Laugier de Tassy (whom Morgan knew in Algiers) in his 1725 *Histoire du Royaume d'Alger*:[29]

> The generality of Christians are so prejudiced against the Turks and other Mahometans that they have scarce language harsh enough to express the horror and contempt in which they hold all of that persuasion. [...] Many

26 *Mahometism Explained*, II, 32.

27 *A Compleat History of the Present seat of war in Africa* (London: W. Mears and J. Stone, 1732), III, 117, 132–134.

28 Nabil Matar, 'Jos.[eph] Morgan and Mohamed Rabadan'. This article also provides references to other studies of Rabadan's work.

29 Jean-Philippe Laugier de Tassy, *Histoire du royaume d'Alger* (Amsterdam: Henri du Sauzet, 1725). This work seems to have been widely read, and was used as a source by Montesquieu, amongst others. See my article in *Christian-Muslim Relations, A Bibliographical History* ed. David Thomas and John Chesworth, vol. 12 (Leiden: Brill, 2020), 545–548.

> persons make no manner of difference between the people of Barbary and real brutes; calling them, simply and absolutely, beasts [...] To such sort of folks the base name of a Turk, Moor, Arab, or other Mahometan is sufficient to inspire them with such opinions. But, I am persuaded, that could those very persons, unknown to themselves, converse with Mussulmans who had no turbans, and were habited after the Christian mode, they would find in them all they can meet with in other people. But if they wear a turban, that article alone suffices to induce them to persist obstinately in their prepossessions.[30]

In his Dedication to Harley in 1723, despite paying lip-service to the prevailing hostility to Islam as a false, inconsistent and incoherent religion, he writes that 'The Chief Motives of my Undertaking the Translation, were, to give my Countrymen more just Ideas of the Notions of the *Mahometans* than they have hitherto received'.[31] This manuscript, copied by Moriscoes in order to teach them the elements of their faith, therefore enables Morgan to give voice to these persecuted people in their own words. He himself is perfectly aware of what he is doing, as appears in his interesting remarks on translation in general, and his translation of this manuscript in particular; while agreeing that sometimes he has provided a paraphrase rather than a translation, he insists that he has nowhere altered the author's true meaning, and 'there are more whole *Pages*, in a manner *Verbatim*, as the *Moor* delivers them, than in any *Translation* extant in our Language'. The changes he has made are in order to make the sense clearer to an English reader, a task for which, according to him, he is well suited as he can guess more easily what a Muslim would say due to his 'intercourse with those people'.[32]

But as he is also aware of the danger of seeming too favourable to this 'false religion', he on occasion insists that he is only the translator; for example, when it comes to Rabadan's criticism of the Trinity (still a particularly sensitive subject in England in the early eighteenth century, when toleration was not accorded to the Unitarians), he adds the following footnote:

> I would, nevertheless, fain have omitted this whole Paragraph, or, at least, have given it a different Turn; but, upon second Thoughts, I consulted some Gentlemen of Worth and Learning, who all told me, *That if I pre-*

---

30 *A Complete History of Algiers* (London: J. Bettenham, 1728), iii–iv.

31 *Mahometism Fully Explained*, I, iv.

32 *Mahometism Explained*, II, ii–iii. This question will be further discussed below.

> *tended to give the World a Translation, I must give the real Meaning of my Author, and not my own, except I did it separately from the Work itself.* But for the Reader's further Satisfaction, I give a faithful Transcript of the whole Paragraph. Videlicet.

After which he transcribes the original Spanish text. Or again, after translating faithfully Rabadan's criticisms of the miracles of Jesus, he defends himself in the following note:

> This Doctrine is indeed, very Anti-Christian, but they are the Words of a Moor, not of a Christian. He writes with the real Sentiments of his honest well-meaning Heart, which he imbibed with his Mother's Milk, and had been handed down to him by the erroneous Traditions of his Forefathers, and to them by their subtle Lawgiver, the Imposter Mahomet.[33]

He also replies to what he considers unfounded and excessive criticisms of Islam and the Prophet in the following terms, being careful to refer to respected and irreproachable specialists to support his point of view:

> Very many grave Grecian, Italian, German, Spanish, and other Writers, have impudently exposed their little-commendable Malice, and impotent Partiality, by painting out Mahomet as something beneath the Human Species, and rendering his Descent still more base and contemptible than they have done his Person and Intellectuals. Now, this really is not *Bonne Guerre*; and all these un-manlike Methods of attacking ones Adversary, have been more than sufficiently exploded, and intirely refuted by some few less ungenerous enemies. Our late worthy Dean Prideaux, and the learned Monsieur Reland, have, in particular, plainly demonstrated, That Mahomet was several Removes from an Ideot, or a Drivler, and, that his Pedigree was not what those Champions make it to have been: Yet, notwithstanding their Gentleman-like Manner of Procedure, in those Points, they represent both himself and his Doctrine in Colours very proper to inspire all, who had the Happiness of being brought up in a Belief more uniform and beautiful, with an Abhorrence of a Persuasion so repugnant to Probability it self, as is certainly the Bulk of what that bold Impostor left behind him, as an unhappy Inheritance to his Proselytes, and their Posterity. I shall, therefore, not give my self any farther Trouble to con-

33 *Mahometism Fully Explained*, I, 214–215.

> fute those partial and indiscrete Historians, nor even to enumerate their Names, but shall enter upon my intended Subject.[34]

Situating Morgan

Such remarks reflect, I think, very well Morgan's conception of what he was trying to do, namely to allow this Other to be heard as faithfully as possible, in his own words, without deformation or censorship, and thus to give him a voice. The stance of Orientalism as criticised since Edward Said's classic work,[35] is that the 'Orientals' cannot speak for themselves. I think that what distinguishes Morgan's enterprise is precisely his desire to allow Rabadan's voice to be heard, following the 'received Maxim' that he quotes at the beginning of his extract from Perez's work: 'A Loser should never be debarred from speaking'.[36] Morgan clearly believed that his translation could allow this Other to speak directly to the English reader and present his belief without deformation. Instead of 'othering' him, Morgan is trying to bring him closer to the English reader. In this, what he was trying to do went further, in my opinion, than other contemporary Europeans who did not espouse the Orientalist view denounced by Said and instead showed sympathy for Muslims.[37] This sympathy often took the form of seeing Islam as a sort of deism or of praising the toleration shown by Muslim governments like that of the Ottoman Empire, often with the aim of criticising the intolerance of Christians, in particular the Catholic Church. Morgan was, like Laugier de Tassy, trying to counter European prejudices against Muslims, and he was doing this by translating the work of this persecuted Morisco.

It is clear throughout his writings that after twenty years spent in North Africa, Morgan felt he understood very well not only the language but also the outlook of the people and wanted to provide the English reading public with a more accurate image of them. He emphasises frequently this first-hand knowledge by using it to add information or correct mistakes. Thus, he presents himself as in a way the authorised mouthpiece of these people, and he certainly does not disappear when letting the Morisco's voice be heard. One could perhaps therefore argue that Morgan is substituting himself for the

34 *Mahometism Explained*, II, v.

35 Edward W. Said, *Orientalism* (London: Routledge, 1978).

36 *Mahometism Explained*, II, 343.

37 See Jürgen Osterhammel, *Unfabling the East. The Enlightenment's Encounter with Asia* (Princeton: Princeton University Press, 2018).

Other and interpreting them, but what emerges from his writings – unusually for Europeans who had spent time in North Africa – is the fact that he seems to have developed a certain emotional identification with the region, despite everything. It is striking, for example, that in one of his histories, he reflects on the destruction of the ancient rich city of Tremezan (modern Tlemcen) by both Turks and Spaniards, in the following words:

> The bare Sight of the wretched, tho' noble and venerable Remains of that once large and beautiful City, have more than once brought Tears to my Eyes; and the doleful stories told me by some of her ancient Citizens have set them flowing.[38]

But neither this sympathy nor his concern for accuracy mean that his books constituted a mere defence of the North Africans, as he very often appears exasperated by the behaviour of both the rulers and the inhabitants of Algiers; and this seems to include all the inhabitants of whatever origin, including the 'renegadoes' (Christians who had converted to Islam), about whom he recounts several unpleasant stories. His condemnation of the Algerines comes out clearly in the Preface to his *History of Algiers*, where he insists that his purpose is to re-establish the truth, not to defend them:

> What a monstrous Load of gross Lies and insufferable Absurdities, as I here and there occasionally observe, does one not fall in with immediately upon laying Hand on any Tract, Treatise, History, Memoir, Relation, Dictionary Geographical, Historical, or Critical, relating to Africa and its Affairs! Yet how many Wide-Acres tell one, scornfully, we have Histories enough of that Country? Why, truly, so we have: But the Question is, have we one, that is not rather Romance than History? Doubtless, those of other Parts are not free from Falsities: But of those I am not so good a Judge.

At the same time, he calls the Algerines 'lawless freebooters' and insists that

> notwithstanding I do them strict Justice, [they] never had a more inveterate Enemy than my self, nor one who more heartily rejoiced at their Disgraces, and wished, nay merely languished for their Destruction, during the whole Time of my Intercourse in their Country.[39]

38 *A Compleat History of the Present seat of war in Africa*, 91.

39 *History of Algiers* 1728, ii–iii.

In this Preface he claims impartiality and his plea, frequently reiterated, is to abandon our prejudices and to judge others by the same standards by which we judge ourselves. He writes: 'Does a people's being, thro' prejudice of education, of a persuasion and religion different from ours, intitle us, merely on that score, to term them savages?'[40]

He is always very careful, when countering prejudices against Muslims, to insist on his simple concern for accuracy and veracity, and to enlist irreproachable specialists like Prideaux in order to forestall any possible criticism that he is showing too much partiality towards the Prophet. Prideaux's polemical life of the Prophet was a defence of Christianity directed against both Islam and freethinkers, in particular irreligious works such as the infamous *Traité des Trois Imposteurs*. Morgan is aware that if he wishes his English readers to listen to the authentic voice of these Muslims as expressed in the manuscripts, he must not seem too partial towards them, hence his reference to the Prophet as an impostor, following the usual practice in England. At one point in his presentation of his translation of Rabadan's work he indicates that he is aware of probable criticisms that such a work, presenting a favourable view of Islam, is dangerous for those whose faith is weak. His reply is that such an argument is worthy of a 'Portuguese Inquisitor', and that on the contrary, this account will make Christians prefer the purity of their own religion, which is more rational than 'that Medley of inconsistent Incoherencies, *Mahomet's Alcoran*'. However, he continues:

> Indeed it contains abundance of good Morality, which might make the best of us blush to see ourselves so much outdone, by an ignorant, tho' subtil Heathen, if I may so call one who detested Idolatry, and taught the worship of the real, the only, and the self existing God, tho' after his own Manner.[41]

This comment indicates once again that, however much he is paying lip-service to Christian views of Islam and its Prophet, he is in fact indicating that it is a religion which is close to the pure Christian religion. This was a view held at the time by many Unitarians, who believed that Christianity had been corrupted by the doctrine of the Trinity amongst others, and that Islam's emphasis on the unity of the deity was close to that of true Christianity.[42] It is not clear whether

---

40 *History of Algiers* 1728, vii.

41 *Mahometism Explained*, II, lvii–lviii.

42 See Justin Champion, *The Pillars of Priestcraft Shaken. The Church of England and its enemies* (Cambridge: Cambridge University Press, 1992), 99–132.

or not Morgan really held such a belief, but what is certain is that his criticism of Idolatry and the intolerance of the Spanish and Portuguese Churches, and of the Catholic Church in general, was sincere and was part of a Protestant vulgate.

Indeed, his other aim in publishing several of his works, including the translation of Rabadan as well as the extracts from the manuscript work by Perez, was to criticise the behaviour of the Spaniards towards the Moriscoes. He provides, for example, a long extract from Perez's work containing a violent and emotional attack on the Spanish Inquisition and its persecution of heretics, which he contrasts with the tolerant behaviour of the Muslims.[43] Morgan precedes this extract with the comment that 'I have, to my best Remembrance, never met with any Writer who has spoken more to the Purpose, or in sharper Terms', informing the reader that if they should find the attack too virulent, then they should reflect that if they had suffered at the hands of the Inquisition, they would probably have used the same language.[44] The extract begins: 'Monsters of Men! says he, Can any Thing, on this side Hell it self, equalize, stand in Competition with, or even, in any wise, approach the Inhumanity, the Hypocrisy, and the monstrous Injustice of your *Infernal Court*'.[45] It describes in bloodthirsty detail the sufferings of the Moriscos in Spain, despite the promises given to them by the Spanish monarchy, and the corruption of the Spanish Church. Morgan's aim is clearly to appeal to the reader's emotions by translating these pathetic descriptions of the sufferings of the Spanish Muslims, written by a member of that community.

This extract is preceded by a text called 'The Case of the Moriscoes or Spanish Moors', which uses a pamphlet by the anti-Catholic polemicist Michael Geddes called *The History and Expulsion of the Moriscoes out of Spain* (1702), but also draws on other Spanish and French sources to provide more detailed historical explanations.[46] Here Morgan recounts in great detail the story of the expulsion of the Moors from Spain and what he calls 'the most inhumane Persecution' lasting many years. Morgan explains that part of the reason for adding this account is to reply to accusations that a manuscript written in Spain dated 1603 – as indicated on the title page of Morgan's publication – must be a fiction because there were no more Moors in Spain at that date. The detailed account is therefore designed to re-establish the historical truth[47]

43 *Mahometism Explained*, II, 343–366.
44 *Mahometism Explained*, II, 343.
45 *Mahometism Explained*, II, 344.
46 *Mahometism Explained*, II, 225–343.
47 *Mahometism Explained*, II, 226–227.

and is part of his self-fashioning as a serious historian. To this effect, he reproduces extracts from a variety of works, both on the Spanish and the Morisco side, beginning with a French Catholic work condemning the inhumanity shown by the Spaniards to both the inhabitants of the New World and the Jews and Muslims in Spain after the Reconquista.[48] He goes on to provide great details of the various stages of Spanish attempts to ensure the conversion of the Spanish Muslims, the latter's resistance, and the increasing persecution up to the final expulsion in 1610. The extracts from works by Geddes and others are interspersed with Morgan's own explanations of particular points, such as a learned discussion of the mixed origin of these Muslim inhabitants or the etymology of the word 'Saracen'. The extracts include a long account by a Spanish monk defending the violence against the Muslims and justifying the actions of the Spanish Catholic monarchs. The author reproduces the decrees against the Moorish practices, customs and language and justifies them by reference to the Moors' duplicity in pretending to be converted to Christianity while remaining secretly Muslims. But the Spanish author also reproduces at length the Moriscoes' reply, defending their customs and attempting to show why the decree was unjust. Morgan then claims that the Spanish author cannot hide the fact that the Spanish did not respect the agreement made when the Moors surrendered, and he provides a description, taken in part from Geddes, of the repression of the Moorish rebellion and the massacres committed by the Spanish. He insists on the way the Spanish reneged on the promises made to the Moors, providing copious details of the different stages of this history. Thus his account steadily moves from apparent objectivity to increasing denunciation of the behaviour of the Spanish, and also of the Pope who freed the Catholic monarchs from their oaths, commenting: 'But what cannot that *infallible* Pastor do among his own Flock!'.[49] His overall attitude can be summed up by the following comment, towards the end of 'The Case of the Moriscoes':

> These *Anti-Christian* Methods of dragging People into the Church by the Hair of the Head, and destroying them *Root* and *Branch*, because they dislike such Usage, is, I believe, peculiar to the Merciful *Church* of *Rome*, and to no other; for it is well known, that even the *Mahometans* themselves never follow that Example.[50]

48 *Mahometism Explained*, II, 231–235.

49 *Mahometism Explained*, II, 283.

50 *Mahometism Explained*, II, 309.

Morgan intersperses his history with personal accounts of individuals he had met in North Africa who had suffered at the hands of the Inquisition, for example a 'New Christian of Jewish extraction' and a Moor born in Tripoli who had been a slave in Cadiz, with a description of the torture suffered by the latter.[51] He also quotes a Morisco he met in Tunis who claimed that if his ancestors had treated the Catholic missionaries the same way as the Swedes did, by forcing them to keep their vows of chastity, then 'we their Posterity, might still have enjoyed our Patrimony in that our Paternal Country'.[52] He even writes that he had been told that 'some of the best Morisco families' in Algiers were secretly imprisoning Spanish priests underground in their homes.[53] The extract from the 1615 manuscript that Morgan was lent by a Spaniard, mentioned above, is also provided to reinforce the condemnation of Catholic intolerance and cruelty, contrasted with the tolerance shown by Muslim governments.[54]

In various places Morgan brings out the parallels between the persecution of the Jews and Muslims in Spain and that of the Huguenots in France, claiming that the French Catholics had become more intolerant since 'the Administration of those zealous Cardinals Richelieu and Mazarine [*sic*]'.[55] Elsewhere he emphasizes the parallels between the oppression of the Moriscoes in Spain and that of the early Christian Church,[56] and when he criticises the 'superstitious and ridiculous ceremonies' of pilgrims to Mecca or the tricks of a *taleb* in Constantine in Algeria, he compares these superstitions to the practices of the Catholic Church. He also quotes a work by an Irish Jesuit published in Lisbon in 1645, inciting the Irish Catholics to destroy all the heretics and drive them out, so that 'our holy Land may be no longer plagued with such a light, changeable, inconstant, barbarous, ignorant, and lawless Generation of People', and to replace the King and the administration with true Catholics subordinate to the Pope.[57] His anti-catholicism, as we can see, prevents him from showing any sympathy to this persecuted Other, the native Irish colonised by the English and chased off their land.

In all of his works, therefore, Morgan appears as a patriotic English Protestant, hostile to the traditional Catholic enemy. He even provides a quotation from Samuel Butler's poem 'Hudibras' claiming that the motive of the Cath-

51 *Mahometism Explained*, II, 237–239.
52 *Mahometism Explained*, II, 241.
53 *Mahometism Explained*, II, 290.
54 *Mahometism Explained*, II, 295–300.
55 *Mahometism Explained*, II, 239.
56 *Mahometism Explained*, II, ix.
57 *Mahometism Explained*, II, 294.

olic Church is not saving souls but acquiring gold.[58] His militant Protestantism comes out very clearly in the critical notes that accompany his translation of the French Trinitarians' account of North Africa, *Voyage pour la redemption des captifs aux royaumes d'Alger et de Tunis* first published in Paris in 1721.[59] This book, by Philémon de La Motte and others, was a French work of Catholic propaganda, dwelling at length on the sufferings of the Christian captives of the Barbary pirates whom it was their task to ransom and bring back to France. Morgan adds numerous footnotes to his translation, correcting the authors' mistakes and adding information, sometimes taken from his own work on Algiers, but above all criticising Popish superstition and accusing the good Fathers of downright lies in their presentation of the situation of the Christian captives in Algiers. For example, he adds the following note to the Trinitarians' account of the Algerines' barbarity:

> We may take it for granted that some Barbary Moors, and these in particular, are as capable of inhumanities as any Spanish Inquisitor, or French Dragoon in Europe. – But might not this good Father have found matter enough, whereon to have exercised his exclamatory talent, much nearer our home? Tho' that, indeed, would not much have answered his purpose.

Or, even more outspoken:

> O Fie, Father! Tho' it is Part of your Function to make a dismal story of Slavery among Infidels (the very Name of which is indeed bad enough) yet you should, methinks, adhere only to the Truth. You came very lately from Marseilles, where you must, or might have seen the Turks, Moors, &c, in a much worse condition than the most unhappy Beylic slave in Algiers; especially since the Algerines have disused Gallies. You likewise must needs have seen or heard, how slaves are treated in Spain, Malta, Genoa, &c. The Beylic slaves of Algiers are (as yourself tells us) at free liberty to play, work for themselves, or steal, three hours before sunset; which latter case many of them take almost with impunity, openly selling whatever they have so acquired, and that not seldom to the very owners, who are sensible that complaining would be fruitless, and that the most satisfactory answer they must expect would be: 'slaves will steal if they

58 *Mahometism Explained*, II, 240.

59 *A Voyage to Barbary, for the Redemption of Captives; performed, in 1720, by the Mathurin-Trinitarian Fathers, François Comelin, Philemon de la Motte, and Jos. Bernard* (London: Charles Corbett, John Brindley, John Jolliffe and Samuel Baker, 1735).

> can; it is your business to be careful of them'. Neither is it less certain, that thousands of Algiers captives live abundantly happier there (want of freedom excepted) than ever they can even hope to do at home; and that very many are excused with a few bastinadoes for crimes, for which they would have suffered the wheel in most parts of Europe, or at least have made their exit in a halter. Therefore I say again, Father, stick to the truth.[60]

He consistently refers in the footnotes to Catholic cruelties (and, as in this passage, to European punishments such as racks and wheels), much worse than those of which the Algerines are accused. He also particularly pours scorn on the 'Popish Pageantries' represented by processions of redeemed captives organised by the Trinitarians all over France in order to extract charity from the faithful:

> Tho' we have no Trinitarian Fathers of Redemption, to roam up and down, and beg money for the relief of our captive brethren among infidels and barbarians, not any who make it their business to parade with them, in pompous, solemn and expensive processions, from one extremity of the Kingdom to the other, again calling out to people for their charity, wherof little falls to the share of those harassed captives, who doubtless had much rather be with their friends, than under this fatigue of masquerading; tho' we have not, I say, any of this pageantry, nevertheless our neighbours have no reason to charge Great Britain with being any way behind-hand with them in this respect. No! these happy islands have their Royal Fathers of Redemption, who, without calling out for the charity of others, are too bountiful to suffer their subjects to languish in bondage, when in their power to procure them a release.[61]

This is part of the 'Postscript by the Translator', in which Morgan also provides lists of the British captives ransomed respectively by King George I and George II.[62] He thus transforms this translation of a work of Catholic propaganda into a patriotic British anti-Catholic vehicle.

---

60 *Several voyages to Barbary containing an Historical and Geographical Account of the Country. With the Hardships, Sufferings, and manner of Redeeming Christian Slaves* …, second ed. corrected (London: Olive Payne, Joseph Duke, Samuel Baker, 1736), 44.

61 *A Voyage to Barbary, for the Redemption of Captives*, 141–142.

62 *A Voyage to Barbary, for the Redemption of Captives*, 141–146.

Morgan on Translation

While it is obvious that Morgan undertook a certain number of translations for purely pecuniary reasons, as he clearly needed the money,[63] others were more likely done for his own specific purposes. This is obviously the case for the translation of Muhamed Rabadan's manuscript, which he seems to have carried out for ideological reasons, and probably also self-promotion. In the case of the work by the French Trinitarian monks, it is difficult to know, as he may have undertaken it himself, or have been commissioned and then put his own slant on it. But whatever the reasons, he seems to have taken the role of the translator seriously and to have reflected on it, because we find, scattered in these publications, reflections on the activity of translation and the relations of languages to each other, as well as the role of the translator as a cultural go-between. We have already seen that he insists on the faithfulness of his translations to the original. In the 'Translator's Advertisement' prefixed to his translation of *The History of the Revolutions in Spain* in 1724, he recognised that some sentences could possibly be criticised by 'grave readers' as 'somewhat too poetical for so grave a History; but I thought it no Business of mine to alter them', except where he has indicated so in notes. While insisting that he has given 'a faithful version, without alteration or diminution' except where indicated in notes, he claims to have corrected some errors in the French edition and to have explained some points. He concludes:

> I do not set up for one of that class whom the judicious Mons. La Motte calls Idolatrous Translators, but have taken the Liberty of introducing a Word or two where-ever I look'd upon the Sense to be defective without them.[64]

He is here referring to Houdar de La Motte (1672–1731), a French playwright and member of the Académie française, who used the expression in his *Discours sur la poësie* (1707), defending his very free translation of Horace and claiming that a 'superstitious respect for the original' does a disservice to genius.[65] Despite Morgan's approval here of La Motte he was in fact far from defending such an extreme position.

63 Probably works such as [L. Ellies Du Pin and J. de Vayrac], *History of the Revolutions in Spain* (London: W. Mears, 1724); or Paul Rapin de Thoyras, *The History of England*, vol. II (London: James Mechell, 1733).

64 *History of the Revolutions in Spain*, v–xviii.

65 *Odes de M. de La Motte de l'Académie Françoise: avec un Discours sur la poésie en général, & sur l'ode en particulier* (Paris: Grégoire Dupuis, 1711), 61.

Morgan makes a similar remark in a passage in his *History of Algiers* in which he criticises abbé de Vertot's *History of the Knights of Malta*, in which the author mangles many names; this has also led his translator into error:

> among the rest, he will needs call the Person I name *Drub-Devil* (from *Cacha-Diablo* the *Spanish* nick-name) *Devil-Driver*; when had he understood *Spanish*, and had perused any of the historians of that nation, who treat of these affairs, he would with far greater propriety, have made it *Bruise-Devil*, which is the word's literal signification: the *French*, indeed, has it *Chasse-Diables*, which is far from being right. But these are blunders into which every translator will, inevitable, fall if he too idolatrously adheres, or rather cleaves to his original; more especially if he happens to be one of those, who had no idea of any of the languages from whence his author translated.

And Morgan goes on to criticise the ignorance of too many translators who 'make horrible havoc' of foreign names, and to insist that a translator should know enough history to be able to correct the original author.[66]

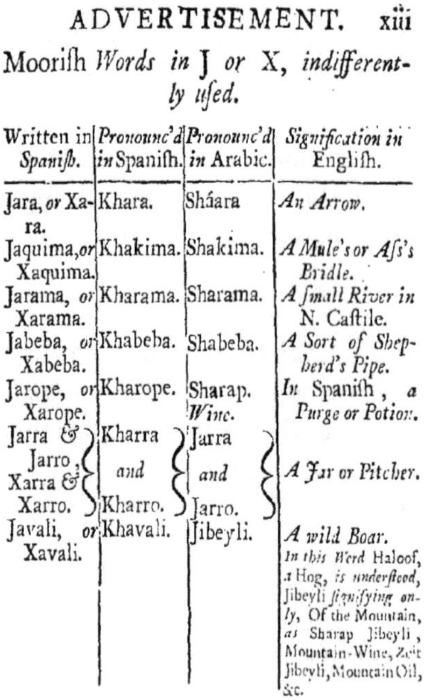
ADVERTISEMENT. xiii

Moorifh *Words in* J *or* X, *indifferently ufed.*

Written in *Spanifh.*	*Pronounc'd in* Spanifh.	*Pronounc'd in* Arabic.	*Signification in* Englifh.
Jara, *or* Xara.	Khara.	Sháara	*An Arrow.*
Jaquima, *or* Xaquima.	Khakima.	Shakima.	*A Mule's or Afs's Bridle.*
Jarama, *or* Xarama.	Kharama.	Sharama.	*A fmall River in* N. Caftile.
Jabeba, *or* Xabeba.	Khabeba.	Shabeba.	*A Sort of Shepherd's Pipe.*
Jarope, *or* Xarope.	Kharope.	Sharap. *Wine.*	*In* Spanifh, *a Purge or Potion.*
Jarra *&* Jarro, Xarra *&* Xarro.	Kharra *and* Kharro.	Jarra *and* Jarro.	*A Jar or Pitcher.*
Javali, *or* Xavali.	Khavali.	Jibeyli.	*A wild Boar.* *In this Word* Haloof, *a* Hog, *is underftood,* Jibeyli *fignifying only,* Of the Mountain, *as* Sharap Jibeyli, Mountain-Wine, Zeit Jibeyli, Mountain Oil, &c.

FIGURE 4.1
Vertot, *History of the Revolutions in Spain* (London: W. Mears and T. Edlin, 1734), xiii

66 *Complete History of Algiers*, 307.

He is indeed particularly exercised about names, and includes in his *History of the Revolutions in Spain* a long passage defending his treatment of proper names, criticising those who anglicise Roman names or Latinise English names. He also insists on his superior knowledge of the transliteration of Moorish names, due to his years spent in North Africa. He devotes particular scorn to the way the Spanish 'mangle' them and includes a learned disquisition on the pronunciation of various letters in different languages, even providing tables of various Moorish words and names, with their Spanish and Arabic pronunciation (see fig. 4.1 above).

He also explains the etymology of various Spanish place names, because 'I have frequently heard Guadalquivir exclaimed against as a most uncouth and barbarous word, whereas when the Etymology is once known there is nothing barbarous in it but the Spanish manner of writing and speaking it'. He provides another much longer table of some of these words, of which this is the beginning:

Written in Spaniſh.	*Pronounc'd in* Spaniſh.	*Pronounc'd in* Arabic.	*Signification in* Engliſh.
Guadalquivir *in* Andaluſia.	Wadalkebeer.	Weit El Kibeer.	*The* Great *River*; *anciently call'd*, Bætis; Weit, *or* Weid, *being no other than* River.
Guadalajara, *or* Guadalaxara, *in* Caſtile	Wadalakhara.	Weit El Hadjar.	*The* Stony *River.*
Guadafión, *in* Caſtile	Wadafión.	Weit El Afión.	*This I take to be the River of* Poppies, *of which* Opium *is made*, *in* Arabick *call'd* Afión.

FIGURE 4.2
Vertot, *History of the Revolutions in Spain* (London: W. Mears and T. Edlin, 1734), xv

These tables show, despite some oddities, Morgan's knowledge of the Arabic spoken in North Africa and his interest in the question of language. In the second volume of his translation of Rabadan's manuscript, he also refers at one point to what he calls the 'Punic Language', the maternal tongue of the North Africans, and its various Idioms,

> each of which has a different Appellation, as *Zinetia, Shilowia, Godemsia, Messababia, Shouwia, Kabeylia,* with some others whose Names are out of my Remembrance, some of which cannot understand each other without Difficulty, whereas between others, there is no more Difference

> than between the *Spanish* and *Portuguese*, or even as near as that of our Counties here.[67]

Morgan includes some remarks on translation at the beginning of the same work. They follow on, to some extent, from his reference to La Motte quoted above, and intervene more directly in the ongoing debates about translation methods, about which he was clearly knowledgeable.[68] He begins these remarks as follows:

> It was an ingenious Thought, of a late celebrated Writer, to compare Translations, more especially those of ancient Poetry, to Vegetables, removed from their own natural, and warmer Climate, and transplanted in a colder. They may be, indeed, *forced* to grow; but few, or rather none, ever arrive to a compleat Perfection. No Person, of any Judgment, will hold, that a literal Version can be possibly good, or, That it can even bear Reading; but I am apt to think, That none, of any Kind whatever, would have been less tolerable than this present Work, had I taken the Method of every where adhering superstitiously to the Author's own Words; were it only on Account of the confinement, under which he laid himself, of cramping all his Lines into Number, Feet, or Measure, even wherein, I cannot say, he had been over regular and exact.[69]

As we can see, he comes down on the side of adaptation rather than strict faithfulness to the text of the original. He even admits that he has produced more of a 'paraphrase' than a translation, but defends this, saying:

> whoever should think it worth while to compare the Copy with the Original, would find, that the Author's real Meaning is not any where altered, and there are more whole Pages, in a manner Verbatim, as the Moor delivers them, than in any Translation extant in our Language.[70]

---

67 *Mahometism Explained*, II, 229.

68 On this, see the classic work by Roger Zuber, *Les belles infidèles et la formation du goût classique en France* (Paris: Armand Colin, 1968), and more recently, amongst others, Fania Oz-Salzburger, 'The Enlightenment in Translation: Regional and European Aspects', *European Review of History: Revue Européenne d'Histoire* 13, no. 3 (2006): 385–409.

69 *Mahometism Explained*, II, ii.

70 *Mahometism Explained*, II, ii. For examples of how Morgan's translation compares with Rabadan's text, see Matar, 'Jos.[eph] Morgan and Mohamed Rabadan', 156–167.

He claims only to have 'taken liberties' where the meaning was obscure and insists that he has 'the Advantage of being more capable of guessing what a Mussulman would say on such, or any other Subjects, than one who has had less Intercourse with those People than my self'. Here again we see him presenting himself as a specialist due to his unique experience of this alien Muslim culture. But he also goes on to make another interesting remark, to the effect that he has tried his best 'at once to keep up to the Standard of the oriental Style, and to render, at the same Time, the Reading not altogether unfamiliar to an English Ear'.[71] This again refers to the translator's permanent dilemma of whether to preserve the specificities of the original style or to make it read easily in the target language. But the reference to Rabadan's 'oriental style' is also interesting, as in general Morgan refers to these people as Africans. The adjective therefore refers to the fact that the author is Muslim, which does make one wonder as to the precise meaning it has for Morgan.

In addition to these general remarks on the faithfulness of the translation, there are also specific comments scattered throughout the work that are designed to show both Morgan's command of the language and culture and his claim to accuracy. For example, he added a note to his translation explaining the precise meaning and etymology of the Arabic word for almsgiving, in which he criticised the usual European translation as 'tithe'. He insisted on the fact that people usually gave much more than one-tenth of their wealth, thus underlining Muslims' generosity.[72] Elsewhere he explained the etymology of the word 'Saracen'.[73] These are clearly part of his self-fashioning as an expert with a unique insight into this culture. In fact, his own voice is very present alongside that of the author he is translating, and he certainly does not correspond to the model of the invisible translator.[74]

Conclusion

There is no doubt that Morgan possessed a deep knowledge of this very different African society on the southern shores of the Mediterranean, and that despite his denials, he had a certain sympathy for its people, if not its rulers, whose arrogance seems to have exasperated him as it did most European vis-

71 *Mahometism Explained*, II, iii.

72 *Mahometism Explained*, II, 189–190.

73 *Mahometism Explained*, II, 228.

74 See Lawrence Venuti, *The Translator's Invisibility: A History of Translation* (London: Routledge, 1995).

itors. As a result, he espoused the role of a go-between helping to make this region better known in his native country.[75] It is also clear that his aim in the translation of Rabadan's work, as in his own works about the region, was to counter European preconceptions not only about these peoples but about Muslims in general and even Islam itself. His was not a completely isolated case, but it is sufficiently rare to be worthy of study. He was also relatively unusual in having obtained his knowledge about Islam and Muslim culture not from university studies but from residence in Muslim countries in North Africa, less common than residence in Istanbul or the Levant. It is not in fact certain that he knew literary Arabic or any Arabic beyond the version spoken in the Maghreb. He was therefore apprehensive of his reception by the English learned public, whence his appeal to recognised specialists, but also his frequent insistence on his unique competence and knowledge, underlined by his frequent reference to his own experience in North Africa and his contacts with the inhabitants of the region.

Nevertheless – unlike most of the other translators studied in this volume – he seems to have failed to obtain the recognition in learned circles that he craved, and his translation of Rabadan's work does not seem to have had much impact and was apparently quickly forgotten. This is probably also linked to its content, less likely to find favour with the circles of scholars for whom the study of Islam often served polemical and proselytising aims. But it may also be the result of the 'uncouth' nature of the manuscript itself, which was apparently not characterised by fine calligraphy. Its author, however much respect he may have enjoyed in North Africa, was not a learned Arabic historian or theologian, but a persecuted Morisco, a refugee who had found asylum in Tunisia. Finally, it was not written in literary Arabic but in a strange mix of Spanish and dialect. There does not seem to have been much that was prestigious about it, despite its unique interest at the time, underlined by Nabil Matar but apparently not perceived so clearly by the contemporary educated public. While Morgan succeeded in facilitating the journey of the manuscript across the Mediterranean to London and exerted his best efforts to allow the voice of a Muslim Morisco to be heard by an English-speaking audience, the lack of echo this voice received attests to the obstacles impeding the journey of such a manuscript. In a way, the difficulties it faced can be compared to the obstacles still facing the movement of people northwards from the southern shores of the Mediterranean.

75 To this extent he can be compared to Antoniutti, for example, studied in the chapter by A. Castagnino.

Morgan's *History of Algiers*, printed in more elegant and expensive in-4° volumes, apparently had a somewhat better reception. It was nevertheless far from achieving the long-lasting success of a work like Thomas Shaw's *Travels, or Observations relating to Several Parts of Barbary and the Levant* published in 1738 and also based on a long stay in Algiers, as pastor to the British factory. But Shaw was university-educated and became professor at the University of Oxford, where his work was published, whereas Morgan had no entry into academic circles and clearly lacked the networks enjoyed by translators such as Laurence Humphrey, Ali Beg / Albertus Bobovius, Annibale Antonini or Bernardino Vestrini, discussed elsewhere in this volume.[76] The fact that Morgan added 'Gent[leman]' after his name on so many of his title pages doubtless betrays his awareness of his lack of university qualifications and desire to present himself as someone of consequence. As an outsider, it was extremely difficult to break into intellectual circles despite his dedications to Harley or Mead, amongst others, and he does not seem to have succeeded in finding a patron. Despite the impressive list of subscribers to his volumes, the fact that he had to publish them himself by subscription indicates reticence on the part of booksellers, who did not believe that they would sell.

In addition, the style in which he writes is not in general that of those serious academics, but is much more familiar, personal and direct, and on occasion he even adds a footnote to correct a statement made earlier in the work. Likewise, his claim to historical accuracy and objectivity, as in the account of the Moriscoes' expulsion from Spain appended to his translation of Rabadan's manuscript, is undermined by the increasingly polemical tone. This tone, more akin to that of a pamphleteer, may have prevented his works being taken seriously, and his polemical intent and desire to allow the viewpoint of the 'African' outsider to be heard conflict with his attempt at self-fashioning as a serious expert. The ambiguity of his self-presentation can also be illustrated by a passage in the Preface to the first edition of his *History of Algiers* in which he refers to criticisms made of his inclusion of Latin quotations from Greek authors. He agrees that he should either have provided good English translations of those Greek authors or, better still, both the original Greek and an English translation. This is indeed what a learned scholar (like Laurence Humphrey, for example) would do, and he admits that quoting a bad Latin translation was the worst solution. His defence is 'that whatever Greek I might have, formerly, learned at school, I have so disused the practice of it', that he now knows practically no Greek at all. He also admits that he did not have an English translation to hand

76 See the chapters by José-Maria Perez Fernandez, Baki Tezcan and Giovanni Lista.

and did not know where to find one, and that he had taken the Latin quotations from a Spanish author.[77] All of this undermines his pretensions to be a serious scholar, and it is difficult to know exactly what to make of it. We hear clearly Morgan's own voice as he speaks directly, even naïvely, to the reader. It contrasts markedly with Shaw's academic tone, which proved much more successful in establishing the Oxford professor's *Observation* as the authoritative work on North Africa. All these factors may help to account for the fact that his works were not successful, that he had to earn a precarious living by hack translation and also tried his hand at political journalism, which does not seem to have been any more successful. He vanished from sight relatively quickly, and the posthumous reputation he achieved was that of a plagiarist.

Nevertheless, his publications (both his own works and his translations) repay study, for their attempt to counter what today we would call Islamophobia and for his determination, particularly in his translation of the manuscript, to allow the voice of these people on the wrong side of the Mediterranean to be heard. Ironically, his own voice all but vanished from historical sight, perhaps because it increasingly represented the position of only a very small minority of Europeans. For Morgan, clearly, translation represented an intervention in contemporary politico-religious questions and was far from being merely a way of earning a living. While allowing the Other's voice to be heard, it was also a way of stating forcefully his own opinions. This was accompanied by a real reflection on translation, on the fundamental questions of how to remain faithful to the source text while adapting the work to the target language and culture. Morgan's travels across the Mediterranean and long residence on its Southern shores made him particularly sensitive to this question and eager to educate the British public about this different culture. His failure is perhaps more a reflection of the preconceptions of the British reading public in the 1730s – and even more as time went on – than of his own shortcomings.

---

77 *History of Algiers*, 1728, xix–xx.

CHAPTER 5

The Gate of Languages Unlocked, or Is It Not?

Ali Bey/Albertus Bobovius and the Limits of Universality in the Seventeenth Century

Baki Tezcan

for Dariusz Kołodziejczyk

This chapter introduces a Turkish translation of a Latin text from the seventeenth century, the *Janua linguarum reserata sive seminarium linguarum et scientiarum omnium* (The Gate of Languages Unlocked, or the Seedbed of All the Languages and Sciences) by John Amos Comenius (1592–1670), a Czech theologian and educator, and its translator, a man who started his life (probably) in the Polish-Lithuanian Commonwealth as Wojciech Bobowski, was forcibly moved to Istanbul, and eventually came to be variously known as Ali Beg (Lord Ali), Ali Ufki (Ali 'of the Horizon', his pen name in Turkish), or Alberto Bobovio/Albertus Bobovius.[1]

Comenius is regarded as one of the first proponents of universal education and the *Janua linguarum reserata*, a textbook that taught Latin and introduced its young readers to several topics of knowledge, came to be translated into more than ten languages, thus embodying the universality of Comenius's goals. The objective of this chapter is to point out the limitations of this universal-

1 Agata Pawlina notes that the Polish equivalents of the Latin names *Albertus* and *Adalbertus* are 'Wojciech' and 'Albert' respectively; see her 'Life and work of Wojciech Bobowski revisited: Bobovius (Ali Ufkî) in Polish source material', *Rocznik Orientalistyczny/Yearbook of Oriental Studies* 77/1 (2024): 26–48 (29); I am grateful to Dr Pawlina for sharing the final draft of her article with me before it was published. There is some debate about how one should pronounce Ali Beg's penname, which could also be vocalised as Ufuki and was indeed vocalised as such by Ali Beg himself, as noted by Judith I. Haug, who is, arguably, the most prominent Ali Beg/Albertus Bobovius specialist today; see her 'Being more than the sum of one's parts: acculturation and biculturality in the life and works of Ali Ufuki', *Archivum Ottomanicum* 33 (2016): 179–190 (179 n. 4). I follow the choice of Cem Behar (see his *Ali Ufkî ve Mezmurlar* [Istanbul: Pan Yayıncılık, 1990], 11–12), mainly because *ufkî* is more correct according to the general principles of Turkish pronunciation of Arabic words with the *-î* ending.

 | DOI:10.1163/9789004731806_007

ity by contextualising the Turkish translation, with special – and extended – attention to the life of its translator, the way in which he has been represented by both his contemporaries and some modern scholars, and his other works, including some of his private notes. I argue that the assumptions of universality in which Comenius operated were very much based in a Christian understanding of the world. This essentially Christian world view was also shared by the many European interlocutors of Comenius's Polish-Ottoman translator, as well as by some modern scholars who have used their testimony while studying his career. In contrast to them, however, Ali Beg/Ali Ufki/Albertus Bobovius was himself a man who, arguably, was much more open to diverse understandings of the world and one's place in it, and was, therefore, better able to identify that which was truly universal among these alternatives.

In order to underline the multi-faceted identity represented by the many names he came to be known by, the most commonly encountered ones of which were Ali Beg and Albertus Bobovius, I refer to the Ottoman Turkish translator of the *Janua linguarum reserata* simply as A.B., so as not to privilege either one of these two names over the other one.[2] In the first part of the chapter I focus on his life, in the second one on his works, with special attention to *Janua linguarum reserata*, and in the third one on the question of universality. The extended attention to the details of A.B.'s life and works, which rendered this chapter longer than others in this volume, is justified by the complexity of A.B.'s identity that enabled him to belong to a rare breed among translators. As someone who felt equally comfortable with Franks and Turks,[3] he was not only able to translate both from Latin to Turkish and from Turkish to western European languages, but also to explain a non-western culture to western Europeans and western Europe to Turkish speakers. Furthermore, A.B.'s mobility was the result of enslavement rather than his own choosing, which was not an exception but almost the rule for Ottoman translators from European languages into Turkish up to the late seventeenth century as discussed in the

2 I was inspired by Haug, who asked whether his signature at the end of his letters in Latin, A.B., could also be read as an abbreviation of 'Ali Bei' see Judith I. Haug, *Ottoman and European Music in ʿAlī Ufuḳī's Compendium, MS Turc 292: Analysis, Interpretation, Cultural Context – Monograph* [Schriften zur Musikwissenschaft aus Münster, Band 25] (Münster: Institut für Musikwissenschaft der Westfälischen Wilhelms-Universität, 2019), 71.

3 On the meanings of these terms, see Baki Tezcan, 'The "Frank" in the Ottoman Eye of 1583', in *The Turk and Islam in the Western Eye (1453–1750): Visual Imagery before Orientalism*, ed. James Harper (Aldershot: Ashgate, 2011), 267–296, esp. 269–272.

introduction to this volume and in the chapter by Giancarlo Casale. Many of his Frankish interlocutors thought he belonged to their land and expected him to move once again, to England, France or the Polish-Lithuanian Commonwealth. Understanding why A.B. did not emigrate from Istanbul requires some attention to different aspects of his life and work, which contributed to the length of this chapter. I should also add that some of the basic facts about him are still far from having been settled in secondary literature, which necessitated a more detailed discussion of certain points in A.B.'s life based on primary sources.

A.B.'s Life

According to Agata Pawlina, who has done the most extensive research into A.B.'s pre-Ottoman past, the Bobowski family originated in Lesser Poland in the fourteenth century, from which its different branches extended their land holdings to Podolia, 'including villages near Lviv', by the seventeenth century.[4] Judging from the fact that A.B. identifies himself as 'Leopolitanus' in two of his works,[5] he probably hailed from the 'Lviv branch' of the Bobowski family.[6] Pawlina even identified a certain 'Albertus Bobowski' who was baptized on 20 March 1606 at the Roman Catholic cathedral in Lviv. However, it remains

4 Agata Pawlina, 'The Creation of an Ottoman Turkish Psalter: Ali Ufki's (Bobovius) *Mezmûriyye* (ca. 1665) and Maciej Rybiński's *Psalmy Dawidowe* (1608)', *The Biblical Annals* 13, no. 4 (2023): 635–660 (640).

5 A.B., *Grammatica Turcico-Latina*, Bodleian Library, Hyde 43, cover page; 'Serrai Enderum', in Cornelio Magni, *Quanto di più curioso e vago ha potuto raccontare ... in viaggi e dimore per la Turchia*, 2 vols. (Parma, 1679–1692), 1: 502–604 (502). The former work has recently been published in a new edition: Ali Ufkî Bey (Alberti Bobovy), *Grammatica Turcico-Latina (1666)*, ed. Fafo Telatar, Yaşar Tokay and Nurdan Besli (Ankara: Grafiker Yayınları, 2020); for an introduction to this work in English, see Hannah Neudecker, 'Wojciech Bobowski and his Turkish grammar (1666): A dragoman and musician at the court of sultan Mehmed IV', DSNELL [*Dutch Studies* published by *Near Eastern Languages and Literatures* in Leiden] 2; no. 2 (1996): 169–192.

6 Pawlina notes, however, that he might also be from the 'Lublin branch ..., especially if we wish to assert that he was raised in the Reformed faith', Pawlina, 'The Creation of an Ottoman Turkish Psalter', 640. I will touch upon A.B.'s connection to Protestant Christianity below; the Swedish ambassador Claes Rålamb was one of his contemporaries who noted this connection; see his 'A relation of a journey to Constantinople', in *A Collection of Voyages and Travels some now first printed from original manuscripts, others now first published in English*, vol. 5, (London: John Walthoe, 1732), 669–716 (703).

uncertain if this record belongs to our protagonist,[7] who is generally believed to have been born around 1610 and could have been Armenian.[8]

We do not know exactly how and when A.B. eventually settled in Istanbul, although various accounts agree that he originally came to Ottoman territory as a slave and was converted to Islam. According to the Swedish ambassador Claes Rålamb, who knew A.B. personally, he was 'taken prisoner in the Venetian war', meaning the Cretan War, which began in 1645. This possibility has been dismissed, however, by Cem Behar, a modern scholar, since Behar believes that A.B. translated *Janua linguarum reserata* in 1643,[9] by which time he must have already been exposed to Turkish for a long time. Behar thus hypothesizes that he must have been captured as a much younger man, entering the Ottoman palace as a page (*iç oğlan*) sometime between 1632 and 1639.[10] Similarly, Pawlina suggests that he could have been enslaved 'in 1633–1634, during the unofficial Polish-Ottoman war (the so-called 'Abaza Pasha war') or in the minor Tatar raids that preceded it'.[11]

It is not prudent to dismiss Rålamb's testimony so quickly, though, as he met A.B. in Istanbul in 1657; therefore, he was quite likely reporting what A.B. told him about his own enslavement.[12] Moreover A.B.'s translation of Comenius's work is actually from the late 1650s and not from 1643,[13] meaning that he would

7 Pawlina, 'Life and work of Wojciech Bobowski revisited', 33; for a reproduction of the baptism record, see Agata Pawlina, *Bobovius – Ali Ufkî: Życie Wojciecha Bobowskiego między faktami a legendą* (Cracow: Wydawnictwo Uniwersytetu Jagiellońskiego, 2023), 24–25.

8 Behar, who provided one of the most detailed biographies of A.B., notes a divergent opinion and explains why 1610 should be preferred; see his *Ali Ufkî ve Mezmurlar*, 10, n. 6. On the possibility of A.B. being an Armenian, see Jean Deny, 'Le MS Marsh 187 de la Bodleienne et Albert Bobowski (Ali Bey)', in *Proceedings of the Twenty-Third International Congress of Orientalists. Cambridge, 21st–28th August 1954*, ed. Denis Sinor (London: Royal Asiatic Society, 1956), 207–208; Il′ja Vladimirovič Zajcev, 'La politique turque de Petro Dorošenko: Documents du fonds de Wojciech Bobowski à la BNF', *Cahiers du monde russe* 50 (2009): 511–532 (512–513); compare Pawlina, *Bobovius – Ali Ufkî*, 21 n. 4.

9 Rålamb, 'A relation', 703; Behar, *Ali Ufkî ve Mezmurlar*, 14.

10 Behar, *Ali Ufkî ve Mezmurlar*, 12–14.

11 Pawlina, 'The Creation of an Ottoman Turkish Psalter', 638; as Behar notes, Polish sources have long asserted that A.B. was enslaved by Crimean Tatars who used to organise raids into the Commonwealth: see Behar, *Ali Ufkî ve Mezmurlar*, 10.

12 Karin Ådahl, 'The twenty paintings depicting the Sultan's Procession', in *The Sultan's Procession: The Swedish embassy to Sultan Mehmed IV in 1657–1658 and the Rålamb paintings*, ed. Karin Ådahl, (Istanbul: Swedish Research Institute in Istanbul, 2007), 74–113 (96).

13 Hannah Neudecker notes the existence of two letters from 1643 in the Paris manuscript of the *Janua linguarum reserata* yet dates the translation to 1658, which is the copy date of the manuscript; see her 'Two Hitherto Little-Studied Turkish Translations by Wojciech Bobowski alias Albertus Bobovius', *Oriens* 45, nos. 3–4 (2017): 330–363 (349–351); com-

indeed have had many years after 1645 to learn Turkish well enough to translate the *Janua linguarum reserata* in the late 1650s. Another of Behar's reasons to dismiss Rålamb's testimony is his assumption that A.B. must have spent nineteen years in the Ottoman palace as a page by the time he was out of the palace in 1657 and met Rålamb, based on the assertion by Paul Rycaut, the secretary of the English ambassador, that Bobovius 'had spent nineteen years in the Ottoman Court'.[14] These nineteen years have usually been understood to refer to the number of years A.B. served as a page in the palace. Yet this would have been an unusually long time for a page to serve in the palace before his *çıkma* ('graduation' from palace service with an appointment to a post outside the palace), especially if he entered the palace at a relatively late age – the only way he could have also gained his well-known competence in several European languages as well as in musical notation, which was not used in Ottoman music.[15] Rålamb's statement that he 'had served ten years for a musician in the seraglio' is, therefore, much more plausible.[16] Rycaut's nineteen years might therefore be inclusive of the time that A.B. served the Ottoman court as a translator for the imperial council (*divan tercümanı*),[17] as well as his years of service as a page in the palace.

Another explanation for the nineteen years could be that A.B. might have continued to serve in the palace as a teacher after his *çıkma* because of his musical skills. He himself notes in his account of the Topkapı Palace, the *Saray-ı Enderun*, that the masters of musical instruments came to the palace from out-

---

pare with Blochet, *Catalogue des manuscrits turcs*, vol. 1, (Paris: Bibliothèque nationale, 1932–1933), 89–90, no. 216. Dr Paul Babinski, who studied the manuscript in question (Bibliothèque nationale de France [BnF hereafter] turc 216), identified the two letters (ff. 428^{b}–429^{a}), one of which bears the date 1643, as Constantijn L'Empereur's foreword and Levinus Warner's preface to the latter's *Compendium historicum eorum quae Muhammedani de Christo et praecipuis aliquot religionis christianae capitibus tradiderunt* (Leiden, 1643). I am grateful to Dr Babinski – whose 'World Literature in Practice: The Orientalist's Manuscript between the Ottoman Empire and Germany' (PhD diss., Princeton University, 2020), includes a section on A.B. and his manuscripts (168–183) – for sharing this identification with me, which suggests that A.B. had borrowed Warner's book from him while the latter was in Istanbul, as I touch upon below.

14 Paul Rycaut, *The Present State of the Ottoman Empire* (London: John Starkey and Henry Brome, 1668), 'The Epistle to the Reader', [2].

15 Behar, *Ali Ufkî ve Mezmurlar*, 10, n. 6; Pawlina, 'Life and work of Wojciech Bobowski revisited', 37.

16 Rålamb, 'A relation', 703.

17 An Ottoman archival document identified by Cengiz Orhonlu names him as a *divan tercümanı* in 1668; see Şükrü Elçin, *Ali Ufkî: Hayatı, Eserleri ve Mecmûa-i Sâz u Söz (tıpkıbasım)* (Istanbul: Kültür Bakanlığı, 1976), v.

side to teach the pages at the *meşkhane*, the music school of the palace.[18] A.B. was so well known as a master of *santur*, a type of dulcimer of Persian origin, that in musical circles he was referred to as 'Ali Beg es-Santuri' (Lord Ali, the *santur* player).[19] His fame as a *santur* player is documented in a chronicle of the Crimean Khanate from the eighteenth century, as I touch upon below.

In any case, if one takes the 'nineteen years of service' at the Ottoman court as inclusive of A.B.'s service as a translator and/or his likely position as a *santur* teacher, then it becomes almost possible to reconcile Rålamb's and Rycaut's statements, as the former added that in 1657 A.B. was 'lately set at liberty' after ten years of service.[20] This would date his original entry into the palace to 1645–1647, while the latter assertion by Rycaut referred to nineteen years when he wrote his book, thus also dating his original entry to the 1640s.[21]

There is another indirect piece of evidence that, if interpreted in a certain way, supports Rålamb's report connecting A.B.'s enslavement to the Venetian war. In a very rare autobiographical statement that Judith I. Haug identified in her meticulous study of private notes that belonged to him, A.B. seems to refer to himself as 'one of the protegés [*çıraks*] of Silāḥdār Yūsuf Pāşā'.[22] This note appears to have been copied from what was originally a petition to Sultan Mehmed IV, when A.B. was still in the palace, but about to leave it at his *çıkma*, quite likely in the 1650s. Silahdar Yusuf Pasha had been executed on the orders of Sultan İbrahim (r. 1640–1648) in January 1646 after he had conquered Chania from the Venetians in 1645.[23] What could A.B. have meant by identifying himself as a protegé (or 'student') of this former grand admiral? How could he have been a *çırak* of Silahdar Yusuf Pasha while he was a page in the palace? I suggest that A.B. might have been taken prisoner at the beginning of the Venetian war in 1645, been presented to the grand admiral in Chania, and then sent by him to the imperial palace as a gift, which would make him a *çırak* of Silahdar

18 A.B./Magni, *Quanto di più curioso e vago ha potuto raccontare*, I: 550; A.B., *Serai Enderum*, trans., Nicolaum Brenner, *Serai Enderum. Das ist: Inwendige Beschaffenheit der Turckischen Kayserl Residentz zu Constantinopoli, die Newe Burgk genant, sampt dero Ordnung und Gebrauchen so von Alberto Bobovio Leopolitano* … (Vienna: Johann Jacob Kürner, 1667), 75; A.B., 'Serai Enderum', Pawlina, trans., in *Bobovius – Ali Ufkî*, 67–180 (127).

19 That is how he is identified in the manuscript of his collection of contemporary music; see Elçin, *Ali Ufkî*, iv, and another source I touch upon in the third section of this chapter.

20 Rålamb, 'A relation', 703.

21 While Rycaut's book was published in 1668 in England, he might have written the account earlier in the 1660s.

22 Haug, *Ottoman and European Music in 'Alī Ufuḳī's Compendium*, 54.

23 For a biography of Silahdar Yusuf Pasha, see Davut Adlığ, 'Şehrîzâde Mehmed Saîd Efendi'nin Tuhfe-i Mustafâviyye fî Beyân-ı Kapudanân-ı Devlet-i Aliyyesi'nin transkripsiyon ve değerlendirilmesi' (PhD diss., Dicle University, 2020), 211–217.

Yusuf Pasha and confirm both Rålamb's testimony and also the statements on the cover page of A.B.'s *Saray-ı Enderun*, which date his palace service to the reigns of İbrahim and Mehmed IV.[24]

There is yet another piece of circumstantial evidence which supports the possibility that A.B. was enslaved in 1645 at the beginning of the Ottoman-Venetian war over Crete. In his *Saray-ı Enderun* he records a plague that cost the lives of more than one hundred pages at the palace, probably in the 1650s, and provides some national or personal identifiers about some of the victims, all thirteen of whom, he notes, had been enslaved during the Venetian war.[25] His closeness to these thirteen men while there were about hundred others who also died might well be the result of their shared bond to the Republic of Venice, which they had all served in Crete, perhaps with A.B.[26]

If he were indeed enslaved at the beginning of the Venetian war in 1645, as Rålamb reported based on his own conversation with him in 1657, A.B. might

24 See A.B./Magni, *Quanto di più curioso e vago ha potuto raccontare*, I: 502; this statement is also found in manuscripts of *Saray-ı Enderun*: see A.B./Pawlina, *Bobovius – Ali Ufkî*, 68, for a reproduction of the relevant page in the British Library ms. Harley 3409. While A.B. certainly refers to things that transpired during the reign of Murad IV (1623–1640) at the palace, he could have heard of these from others who had been there then and were still at the palace in 1645.

25 While he is not providing an exact date for this plague, he notes that it followed a 'Küçük Bayram', or 'picciola Pasqua', which took place in summer months; A.B./Magni, *Quanto di più curioso e vago ha potuto raccontare*, I: 525–526; A.B./Pawlina, *Bobovius – Ali Ufkî*, 95–97. As pointed out by Pawlina, 'Küçük Bayram' refers to Eid al-Fitr and not to Eid al-Adha as the German translator of *Saray-ı Enderun* understood it; compare Pawlina, *Bobovius – Ali Ufkî*, 95, n. 135; A.B./Brenner, *Serai Enderum*, 32–33: '*Kutchiuk Bajeram*, oder kleinen Ostern (das *Sacrifacium Abrahami* von ihnen genannt)'. Brenner specifies further that the 'Küçük Bayram' that preceded the plague was in June, which, if correct, would mean that the plague must have happened in 1659 or 1660. If one were to add the months of July and August to the likely times of this plague as suggested by the Italian version of *Saray-ı Enderun*, where it reads 'giorni caniculari', or 'dog days' (A.B./Magni, I: 525), the likely years would expand to 1653–1660.

26 The French version of A.B.'s *Saray-ı Enderun* by the French ambassador Pierre de Girardin in 1686 – which has several differences when compared to the Italian original or the German translation – does not include the identification of these thirteen victims, but the ambassador relates that the plague happened at the palace in Edirne, a detail that is not present in the Italian or German versions; see *Topkapi: Relation du serail du Grand Seigneur*, ed. Annie Berthier and Stéphane Yerasimos (Arles: Actes Sud, 1999), 58; for an English translation of this French version, see Carol G. Fisher and Alan W. Fisher, '*Topkapı Sarayı* in the Mid-Seventeenth Century: Bobovi's Description', *Archivum Ottomanicum* 10 (1985): 5–81 (35); Fishers also interpreted 'la petite pâque' as Eid al-Adha, most probably because of the sheep that were slaughtered and the fact that Girardin interjected an explanation of Eid al-Adha that is not present in A.B.'s Italian text.

have previously been in Venetian service, which would explain his close affinity with Italian. Judith Haug – who arguably spent more time than anyone with the writings of A.B., including his personal notes that include musical, medical, political reflections, as well as miscellenia of all sorts – states that he 'thought in Italian ... While notes on music are invariably formulated in a mix of Italian and Ottoman Turkish or entirely in Italian, Polish texts rarely appear'. She also finds evidence for his Italianate upbringing in the 'Italian lute tablature' that A.B. used, adapting it for different Ottoman instruments.[27] Both Haug and Pawlina suggest that he might have had a Jesuit education. Pawlina specifically points to 'the Jesuit College (*Kolegium Jezuickie*) in Lviv, established in 1608, with a music boarding school (*bursa muzyczna*) operating from 1615'. However, she also notes that 'lists of teachers and alumni of schools, colleges and seminars run by the order' did not include the name Bobowski.[28] Of course, A.B. could have learned Italian at another school or from a private tutor, too. Yet it is interesting to observe that Pier Mattia Tommasino, who studied A.B.'s *Saray-ı Enderun* from a linguistic perspective as an Italian work written by a non-Italian, notes that his 'written idiolect presents a large number of lexical items from Venetian dialect',[29] once again connecting A.B. to Venice. Thus, he might well have been one of the many Europeans who served Venice in its colony of Crete.

Whether he was enslaved by Crimean Tatars in the 1630s or by the Ottoman forces that conquered Chania in 1645, A.B. became a page in his twenties (or thirties), which might appear as an old age for a page at the Ottoman palace but was becoming common in the seventeenth century,[30] and he certainly

27 Haug, *Ottoman and European Music in ʿAlī Ufuḳī's Compendium*, 13, 84, 149. The few examples of A.B.'s written material in Polish have been closely studied by Agnieszka Ayşen Kaim, 'In search of a Polish layer under the Ottoman identity of Albert Bobowski alias Ali Ufkî: From being a "Polish captive" to becoming an *in-between*', in *Becoming Ottoman: Converts, Renegades and Competing Loyalties in the Early Modern and Modern Ages*, ed. Yavuz Köse, Petr Kucera and Tobias Völker (London: I.B. Tauris, 2025), 77–98.

28 Pawlina, 'Life and work of Wojciech Bobowski revisited', 36; Haug, *Ottoman and European Music in ʿAlī Ufuḳī's Compendium*, 85; Pawlina, *Bobovius – Ali Ufkî*, 32.

29 Pier Mattia Tommasino, 'Travelling East, Writing in Italian: Literature of European Travel to the Ottoman Empire Written in Italian (16th and 17th Centuries)', *Philological Encounters* 2 (2017): 28–51 (44).

30 The average age of boys levied from among the Christian population of the empire was increasing already at the beginning of the seventeenth century. Gülay Yılmaz notes that '42 percent of the boys in the [1603/04] register were 18 years old and above', see her 'The Devshirme System and the Levied Children of Bursa in 1603–1604 A.D.', *Belleten* 79 (2015): 901–931 (924). Among enslaved Europeans, one could find older pages. A.B., for instance, refers to a 'captain' named Narciso Marchetti who had been enslaved during the Cretan War, while he is talking about his fellow pages who died in the plague (mentioned

spent some time at the Ottoman palaces in Edirne and Istanbul.[31] A.B.'s exit from the palace is dated with less ambiguity than his entry. In 1657, when he met Rålamb, he was no longer at the palace. His *çıkma* is usually associated with a trip to Egypt in the company of a high-ranking Ottoman officer.[32] In his *Saray-ı Enderun*, A.B. identifies this officer as Galatalı Firari Kara Mustafa Pasha, who was indeed appointed Governor of Egypt in 1656 and dismissed in 1657 by Köprülü Mehmed Pasha (grand vizier, 1656–1661), who, according to A.B., saw him as a rival because he was close to both Sultan Mehmed IV and to the Queen Mother Hatice Turhan. Apparently, Mustafa Pasha subsequently went into hiding until Köprülü died in 1661, hence, the epithet *firârî*, or the 'fugitive'.[33] Thus, if A.B. went to Egypt in the company of an Ottoman grandee who was instrumental in his graduation from the palace school, it must have been in 1656 in the company of Mustafa Pasha, and he must have returned to Istanbul in 1657 while his patron was still in hiding, as he met Rålamb there in that year.

However, it is quite likely that A.B.'s *çıkma* was actually a bit before 1656 and not necessarily related to his trip to Egypt, because his first translation commissioned by a western European patron is dated 1654, predating the appointment of Mustafa Pasha to the governorship of Egypt. Hannah Neudecker suggests that this translation must have been commissioned by Isaac Basire, who served Charles I as chaplain for about five years in the 1640s before he was imprisoned

above) at the palace; A.B./Magni, *Quanto di più curioso e vago ha potuto raccontare*, 1:526. Someone who reached the rank of 'captain' would have been at least in his late twenties, if not in his thirties.

31 A.B. notes his arrival from the palace in Edirne to the one in Istanbul in his account of the palace: see A.B./Magni, *Quanto di più curioso e vago ha potuto raccontare*, 1:585. It is usually assumed that A.B. was at the Edirne palace during the reign of Murad IV and moved to Istanbul during the reign of Ibrahim. If we rely on A.B.'s testimony to Rålamb in determining his date of enslavement, we would have to change this chronology and place him in Edirne in the second half of the 1640s.

32 See, for instance, Thomas Hyde's preface to Albertus Bobovius, 'A Treatise Concerning the Turkish Liturgy, the Pilgrimage to Mecca, Circumcision, Visiting the Sick, etc.', ed. Thomas Hyde in *Four Treatises concerning the Doctrine, Discipline and Worship of the Mahometans* (London: J. Darby, 1712), 103–150 (106).

33 I could not find a reference to Mustafa Pasha in Magni's edition of the *Saray-ı Enderun*; while the French version refers to him, it does not identify him as A.B.'s patron who secured his exit from the palace (see Berthier and Yerasimos, *Topkapi*, 113); the German translation, however, does (see A.B./Brenner, *Serai Enderum*, 110); for more details on Mustafa Pasha's career, see Berthier and Yerasimos, *Topkapi*, 157–158, n. 321; and Adlığ, 'Şehrîzâde Mehmed Saîd', 258–262. A.B. also identifies Mustafa Pasha as his patron in at least one of the manuscripts of the *Saray-ı Enderun* in Italian: see Pawlina, *Bobovius – Ali Ufkî*, 146.

and then exiled for his support of the king during the English Civil War.[34] His exile brought Basire to Aleppo, Jerusalem and, eventually, Constantinople. Basire had a habit of translating the catechism of the Church of England, and of gifting these translations to local church leaders during his travels.[35] He is known to have translated the text himself into Greek while in Zante, and he later also presented an Arabic translation to the Patriarch of Antioch while in Aleppo in 1652. While a later source discovered by Neudecker suggests that Basire also commissioned A.B. to do a Turkish translation, Colin Brennen states that Basire instead 'arranged for the English ambassador in Constantinople to have it translated into Turkish, so that he could send copies to the Armenian bishops he had met in Mesopotamia'.[36] Basire seems to confirm the ambassador's role in commissioning this translation in a letter he wrote in 1653: 'I am now intending to send our catechism in Turkish to some of their Bishops, Armenians most of them. This Turkish translation is procured by the good care of Sir Thomas Bendyshe, ambassadour here'.[37]

A.B., on the other hand, in a letter composed in 1666, thanked Basire for a recommendation that secured his entry into Thomas Bendish's service.[38] Perhaps Basire liked the translation, which was originally commissioned by Bendish, and his approval helped the latter to continue working for the English embassy. In any case, it is known that he 'lived in the English embassador's house' for some time in 1657.[39] A.B. eventually dedicated a copy of his translation of the catechism from Latin to Turkish – which is arranged in two columns on each page with the Latin text on the left and the Turkish on the right – to John Bendish, the son of the ambassador, with references to learning Turk-

34 Hannah Neudecker, 'From Istanbul to London? Albertus Bobovius' Appeal to Isaac Basire', in *The Republic of Letters and the Levant*, ed. Alastair Hamilton, Maurits van den Boogert and Bart Westerweel (Leiden: Brill, 2005), 175–196 (182). I should also add that A.B. does not mention anything about Egypt in connection with Galatalı Firari Kara Mustafa Pasha in the *Saray-ı Enderun*.

35 Colin Brennen, 'The life and times of Isaac Basire' (PhD diss., University of Durham, 1987), 75–76, cited by Neudecker, 'Two Hitherto Little-Studied Turkish Translations', 340–341, n. 38.

36 Brennen, 'The life and times', 83.

37 From Basire to Richard Brown (the English ambassador in Paris), written in Pera, July 20, 1653, in W.N. Darnell, *The Correspondence of Isaac Basire, …, in the Reigns of Charles I and Charles II, with a Memoir of His Life* (London: John Murray, 1831), 115–120 (117); see also his undated letter to 'Ant. Leger, a Protestant minister in Geneva', in Darnell, 121–126 (124): 'à present ie le fais traduire en Turquesque, pour l'envoyer aux Créthiens de Mesopotamie, auquels ie l'ay promis quand ie l'ay visité il y a quelque temps'.

38 Neudecker, 'From Istanbul to London', 193.

39 Rålamb, 'A relation', 703.

ish. This might suggest that A.B. was employed as a teacher of Turkish for the ambassador's son.[40] Also, it is unclear how Basire would have come to know A.B. at this time, since he had just come to Istanbul from Aleppo. Thomas Bendish, however, had been in Istanbul since September 1647, when he began his appointment as the English ambassador. Therefore, it is more reasonable to assume that it was Bendish who knew A.B. first and commissioned the translation.

In any event, A.B. could not have worked as a Turkish language teacher to the son of the English ambassador or received a commission from the ambassador to translate the catechism of the Church of England, while he was still a page in the imperial palace. Therefore, it might be safest to assume that his *çıkma* happened, at the latest, in 1653, when he was commissioned by Bendish to translate the catechism of the Church of England. This date would also explain how he became a teacher for Franciscus à Mesgnien Meninski (d. 1698), a linguist and diplomat, who would go on to publish the monumental *Thesaurus Linguarum Orientalium*, a multilingual dictionary of Turkish, Arabic and Persian in 1680. According to Behar, Meninski arrived in Istanbul in the company of the Polish ambassador in 1652 and studied Turkish with A.B. between 1652 and 1654.[41]

But how could A.B. have been introduced to the English ambassador or become the Turkish teacher of a diplomat who had recently arrived from the Polish-Lithuanian Commonwealth, in the first place? His petition to Mehmed IV the week before his *çıkma* offers a clue:

> Your auspicious and felicitous and merciful highness, my Sultan, be in good health. I have the following petition, my Sultan's highness. This your slave, being one of the proteges of Silāḥdār Yūsuf Pāşā – who has been received in God's mercy and whose sins have been forgiven – being in the process of *çıkma* …, but having no place to go when we leave next week; [but] we have knowledge of many various languages, being able to serve in any capacity as interpreter to the ambassadors, receiving and giving answers and being capable of reading and writing letters; [thus] it is our humble request from my merciful Sultan's highness's excellent kindness

40 Neudecker, 'Two Hitherto Little-Studied Turkish Translations', 338.

41 Cem Behar, 'Wojciech Bobowski (Ali Ufkî): Hayatı ve Eserleri (1610?–1675)', in Cem Behar, *Musıkiden Müziğe: Osmanlı/Türk Müziği: Gelenek ve Modernlik* (*Makaleler-Kaynaklar-Metinler*) (Istanbul: Yapı Kredi Yayınları, 2005), 17–55 (26); Cem Behar, *Saklı Mecmua: Ali Ufkî'nin Bibliothèque National de France'taki* [*Turc 292*] *Yazması* (Istanbul: Yapı Kredi Yayınları, 2008), 35.

> that this slave of slaves may appear worthy to find approval to busy himself in service with righteous prayer as a servant to my Sultan until he dies, that is, bound by an enduring decree …; I am my kind Sultan's highness's bondsman, Ali.[42]

Thus, A.B. might have been directly appointed as an interpreter to the English embassy and the Polish diplomatic legation by the Ottoman administration. His residence at the English embassy may be related to this appointment as well.[43] Alternatively, if A.B. failed to secure the appointment he sought before his *çıkma*, he might have tried to create a position for himself by getting closer to foreign embassies in the imperial capital. Either way, he probably needed some additional income, especially after 1656 when Köprülü Mehmed Pasha became grand vizier.

To understand why A.B. would have needed additional income during Köprülü's grand vizierate, we have to remember that when he left the Ottoman palace, A.B. held the rank of a *sipahi*,[44] that is, a member of the central cavalry corps.[45] At that time, most of the imperial cavaliers were supplementing their depreciating income in various commercial and financial enterprises or as tax collectors.[46] Having access to such opportunities, however, was closely related to one's network and, as noted above, A.B.'s Ottoman patron, who had secured his graduation from the palace, was at that time in hiding. A.B. was not fond of the grand vizier Köprülü Mehmed Pasha, who was hostile to his patron; he referred to him as a tyrant not only in his *Saray-ı Enderun*, but also in his private notes, and called him a murderer elsewhere.[47] Thus, feeling insec-

42 Haug, *Ottoman and European Music in ʿAlī Ufuḳī's Compendium*, 54.

43 Bertold Spuler lists him as the chief interpreter of the Sublime Porte around 1650; see his 'Die Europäische Diplomatie in Konstantinopel bis zum Frieden von Belgrad (1739)', *Jahrbücher für Kultur und Geschichte der Slaven* 11 (1935): 53–115, 171–222, 313–366 (175). While this date is rather too early for A.B., who might have held this position later in his career as will be touched upon below, it might be the right date for the beginning of his career as an interpreter in Istanbul.

44 For Hezarfen Hüseyin's (d. 1691) statement on this, see Ahmet Nas, 'Hezârfen Hüseyin Efendi'nin 'Tenkîhü't-tevârih-i Mülûk' isimli eserinin tahlili ve metin tenkidi' (M.A. diss., Erzincan Binali Yıldırım University, 2019), 491.

45 Rålamb, 'A relation', 703, states that he 'received *spahi*'s pay', which would suggest a *kapıkulu sipahi* rather than a timariot.

46 Baki Tezcan, *The Second Ottoman Empire: Political and Social Transformation in the Early Modern World* (New York: Cambridge University Press, 2010), 184–185.

47 A.B./Brenner, *Serai Enderum*, 97; Pawlina, *Bobovius – Ali Ufkî*, 146; Behar, *Saklı Mecmua*, 38, 109; the source of the political text excerpt among A.B.'s private notes in which he inserted Köprülü's name as a tyrant was identified by Haug, 'Being More than the Sum

ure and facing limited career prospects, he may have turned to the foreign embassies as the only obvious opportunities for work in Istanbul. If he had indeed been serving Venice in Crete before his enslavement, he might already have been familiar with the European diplomatic corps and translation-related professional opportunities.

Whether he was appointed by the Ottoman administration as an interpreter (and wished to supplement his income by receiving additional assignments from the diplomats he was working with) or was trying to create positions for himself by securing assignments or commissions, A.B. needed to build trust with European diplomats he interacted with. Yet he was a renegade and, in the eyes of most European diplomats who were stationed in Istanbul, renegades were, as the Venetian *bailo* Matteo Zane wrote in 1594, 'the most arrogant and evil men imaginable, having lost, along with their true faith, all humanity'.[48] As Tobias Graf explains, 'that renegades were even crueller and more given to worldly pleasures than "the Turks" ... appears as a stock trope in Christian-European writings on the Ottoman Empire'.[49] That this stock trope was alive and well in the 1650s, when A.B. was making his first contacts with European diplomats in Istanbul, is evidenced by Jakab Harsányi Nagy, a Hungarian diplomat, who disparaged renegades and expressed his preference for born Muslims over them.[50] Thus, if A.B. wanted to gain the trust of his potential European patrons, he had to prove himself to them. I would suggest that we need to interpret some of the statements attributed to him in this light. Take, for instance, Rålamb's explanation for A.B.'s residence at the English embassy: '[A.B.] lived in the English embassador's house, in hopes of getting, by his help, out of Turky, and among Christians again, being in his heart still addicted to his former religion of the reformed profession.'[51]

---

of One's Parts', 186, n. 39; see also Virgilio Malvezzi, *Il ritratto del privato politico christiano* (Macerata, 1635), 42–43. I would like to acknowledge the help I received from Amanda Batarseh and Leonardo Giorgetti, who were both graduate students in Comparative Literature at UC Davis back in 2014 when I started working on this project, in transcribing and translating this text. For A.B.'s depiction of Köprülü as a murderer, see Elçin, *Ali Ufkî*, 220; M. Hakan Cevher, 'Ali Ufkî Bey ve Hâzâ Mecmû'a-i Sâz ü Söz (Transkripsiyon, İnceleme)' (PhD diss., Ege Üniversitesi, 1995), 119.

48 Tobias P. Graf, *The Sultan's Renegades: Christian-European Converts to Islam and the Making of the Ottoman Elite, 1575–1610* (Oxford: Oxford University Press, 2017), 29.

49 Graf, *The Sultan's Renegades*, 32.

50 Behar, 'Wojciech Bobowski', 31; while Behar believes that Nagy learned Turkish from A.B., Gábor Kármán states that they may not have met; see his *A Seventeenth-Century Odyssey in East Central Europe: The Life of Jakab Harsányi Nagy* (Leiden: Brill, 2016), 210.

51 Rålamb, 'A relation', 703.

Similarly, Thomas Hyde, who was the keeper of the Bodleian Library and taught at Oxford, wrote in the preface to A.B.'s *Treatise concerning Turkish Liturgy*:

> But it is most of all to be lamented, that Death snatch'd him away, before he had return'd to Christianity; which he earnestly long'd for from his very Soul, wishing he could get his Bread some honest way in England among Christians, and leave the Fellowship of Infidels.[52]

Hyde had never met A.B., who had written the Latin version of his *Treatise* for Thomas Smith, an English professor of theology, who then conveyed the treatise to Hyde when he came back from Istanbul. So, it must have been Smith who reported A.B.'s longing for a life 'in England among Christians' to Hyde. Smith was in Istanbul as chaplain to Sir Daniel Harvey, the English Ambassador, from 1668 to 1671. Therefore, one gets the impression that A.B. told Smith that he longed to move to England, which he might very well have done as it would help him build a relationship of trust in an environment where converts to Islam were not trusted.

The assertion that A.B. wanted to return to Christendom was reproduced in the twenty-first century by some scholars studying his life and works. I will discuss these modern authors later in this chapter. Here, though, I would like to dwell on the contemporary reports a bit more. Might A.B. have indeed hoped to leave the Ottoman Empire? And was he really a Protestant? The answer to both these questions could have been yes. Yet even though A.B. was very close to the English embassy as well as the Dutch one, as will be discussed shortly, he does not seem to have ever attempted to leave the Ottoman Empire after his *çıkma*. Had he wanted to, he certainly would have had opportunities to do so. For example, his student Meninski, who became the deputy ambassador for the Polish-Lithuanian Commonwealth in Istanbul in the second half of the 1650s, could certainly have facilitated a return to his homeland.

In fact, there is another report, by a French monk of the Capuchin order Robert de Dreux, who met A.B. in 1669 in Istanbul, right around the time that A.B. is supposed to have shared his wish to move to England with Smith, that actually gives us an opportunity to observe his reaction when offered a way out of the Ottoman Empire. De Dreux states that Pierre de Girardin, a Frenchman who was to become French ambassador in the late 1680s and who penned the French version of A.B.'s *Saray-ı Enderun*, invited him to France 'where he could

52 Bobovius, 'A Treatise', 105–106.

freely profess the true religion', promising both a pension and good appointments if he did so. A.B. apparently responded by saying that he 'would gladly take this decision' if it were not for his wife whom he could not abandon.[53]

Although we do not know the identity of A.B.'s wife, the fact that he refers to an in-law in Egypt in a personal note from 1671 confirms that he was married.[54] As pointed out by Rifaʿat Abou-El-Haj, Ottoman officials with a palace background usually married former palace servants who were once slaves employed at the *harem*, that is, the part of the palace where the sultan's family and concubines resided.[55] Thus it is quite likely that A.B.'s wife was a former Christian like himself whom he could have taken to France or another European country if he had really wanted.

When pressured by a countryman of his about why he was not returning home, A.B. does not seem to have bothered to make up an excuse about his wife. Teofil Rutka (1622–1700), a Jesuit who accompanied the Polish envoy Jan Szumowski to Constantinople, returned to the Jesuit College of Lviv in 1661 (the year the college became an academy with a royal charter from John II Casimir) with a copy of *al-Hidaya*, a Hanafi legal manual by al-Marghinani (d. 1197), which was sent by A.B. 'so that the dogmas of the Mohammedans might be better refuted ... and that studies of Oriental languages and recondite sciences might be promoted'. Rutka noted ten years later that even though A.B.

> intended to return to his homeland and wished to serve it with his expertise in Oriental languages, of which he had decided to become professor at the Academy of Lviv, ... captivated by the pleasantness of the city of Constantinople, as he himself confessed to [Rutka], he postponed his return to his homeland.[56]

---

53 Haug, *Ottoman and European Music in ʿAlī Ufuḳī's Compendium*, 69.

54 Behar, 'Wojciech Bobowski', 29.

55 Rifaʿat Abou-El-Haj, 'The Role of Women in the Ottoman Empire: How the *khassa* Reproduces the *khassa* – Elite Reproduction in Sixteenth Century Ottoman Jerusalem', in *Festschrift Hans Georg Majer: Frauen, Bilder und Gelehrte – Studien zu Gesellschaft und Künsten im Osmanischen Reich / Arts, Women and Scholars – Studies in Ottoman Society and Culture*, ed. Sabine Prätor and Christoph K. Neumann, vol. 1 (Istanbul: Simurg, 2002), 185–193.

56 For the full text of the inscriptions (in Latin) by A.B. and Rutka in the first and second volumes of this Arabic manuscript, see Gustav Flügel, *Die arabischen, persischen und türkischen Handschriften der Kaiserlich-Königlichen Hofbibliothek zu Wien*, vol. 3 (Vienna: K.K. Orientalische Akademie in Wien, 1865–1867), 204, n. 1. I would like to thank my colleague Daniel Stolzenberg for editing the English translation that I obtained using artificial intelligence (Claude).

Thus, once again, we witness A.B. creating the impression of wishing to move back to Christendom where his skills would secure a comfortable life for him and yet not making the move. If A.B. were voting with his feet, he was clearly choosing to stay in Istanbul. I should also add that while A.B. talked about refuting Islam in his dedicatory note to the manuscript he sent to his home town, his own writings on Islam were actually quite sympathetic for the European standards of the period, as will be discussed below.[57]

As for Rålamb's suggestion that A.B. had been a Protestant, I would like to make a cautionary remark. If A.B. is indeed the above-mentioned Albertus Bobowski in the baptism record that Pawlina discovered in the Catholic church archives in Lviv, then he had most probably been a Catholic before converting to Islam. Whether he was Catholic or Protestant, however, A.B. might have identified his former religious allegiance differently in different contexts. As I pointed out in my study of the *Treatise on Islam* by İbrahim Müteferrika (d. 1747), another renegade intellectual from a later generation, it is possible to find several conflicting testimonies about his religious allegiance before his conversion to Islam by Europeans who met him in Istanbul. I suggested that the conflict between these testimonies might well be of İbrahim Müteferrika's own making, as he seemed to claim Lutheran origins when he talked to a Lutheran, and Catholic origins when he conversed with a Catholic. After all, a shared confession, even if it was in the past, provided a certain bond.[58] Thus, while interacting with Swedish or English diplomats, A.B. might have similarly found it advantageous to claim that he had been a Protestant, regardless of the truth.

A.B. eventually moved out of the English embassy and, at least by the second half of the 1660s, lived in his own rented lodgings.[59] By this period, he was work-

57 Whether or not A.B. really meant Rutka to refute Islam, the latter indeed wrote a polemic against Islam in Latin, *Gladius contra Turcas* (Lviv, 1679), which he translated later into Polish; see Norbert Frejek, 'Teofil Rutka', in *Christian-Muslim Relations: A Bibliographical History*, vol. 8: *Northern and Eastern Europe (1600–1700)*, ed. David Thomas and John Chesworth (Leiden: Brill, 2016), 807–813.

58 There are also reports about him having been Socinian or Jewish; see Baki Tezcan, 'İbrâhîm Müteferrika ve *Risâle-i İslâmiyye*', in *Kitaplara Vakfedilen Bir Ömre Tuhfe: İsmail E. Erünsal'a Armağan*, ed. Hatice Aynur, Bilgin Aydın and Mustafa Birol Ülker, vol. 1 (Istanbul: Ülke Yayınları, 2014), 515–556 (546–548); for an updated (and summarised) version of this article in English, see 'Secularist Anxieties Meet Evangelical Ones in Modern Turkish Historiography: İbrahim Müteferrika and the *Risale-i İslamiye*', in *Transforming Empire: The Ottomans from the Mediterranean to the Indian Ocean – Essays in Honor of Linda Darling*, ed. Serpil Atamaz, Onur İnal and Alexander Schweig (Leiden: Brill, 2024), 45–69.

59 Behar, 'Wojciech Bobowski', 28–29; see also Haug, *Ottoman and European Music in 'Alī Ufuḳī's Compendium*, 69.

ing as an interpreter at the Ottoman Imperial Council (*divan tercümanı*), as evidenced by archival records from 1668.[60] Hezarfen Hüseyin refers to him as holding the rank of 'second interpreter' in his world historical chronicle, written in the early 1670s with some help from A.B. in translating sources on Roman history.[61] Joseph von Hammer suggests that he served as the chief interpreter for a very short term in between the tenures of Panagiotis Nikousios (d. 1673) and Alexander Mavrocordatos (d. 1709), which is difficult to substantiate as the latter seems to have succeeded the former upon his death in 1673.[62] Moreover, if one were to follow Cornelio Magni in determining A.B.'s year of death, he could not have been alive in 1673. Like so many other things about his life, even the date of A.B.'s death is disputed in scholarship. Behar takes 1675 as a plausible year;[63] however, as Barnette Miller points out, Cornelio Magni states in a letter written from Istanbul on 14 October 1672 that A.B. had died a few months earlier, which does not seem to be contradicted by any other primary source that I am aware of.[64]

60 See note 17 above; see also Gunnar Hering, 'Panagiotis Nikousios als Dragoman der kaiserlichen Gesandtschaft in Konstantinopel', *Jahrbuch der österreichischen Byzantinistik* 44 (1994): 143–178 (160).

61 Nas, 'Hezârfen Hüseyin Efendi'nin', 491; for the dating of this chronicle, see Hüseyin G. Yurdaydın, *İslâm Tarihi Dersleri* (Ankara: Ankara Üniversitesi İlahiyat Fakültesi Yayınları, 1971), 135.

62 Joseph von Hammer, *Geschichte des osmanischen Reiches*, 10 vols (Pest: C.A. Hartleben, 1827–1835), VI: 307; the source von Hammer is citing simply notes that Mavrocordatos succeeded Nikousios; see Édouard de la Croix, *Mémoires Du Sieur De La Croix, Cy-Devant Secrétaire de l'Ambassade de Constantinople, Contenans Diverses Relations très-curieuses de l'Empire Othoman*, vol. 2 (Paris: Claude Barbin, 1684), 4; von Hammer also refers to reports by Kindsberg; however, the published version of these reports do not seem to mention A.B.: see Janusz Woliński, 'Wojna polsko-turecka 1672–1676 w świetle relacji rezydentów austriackich w Turcji', *Studia i Materiały do Historii Wojskowości* 7, no. 2 (1961): 322–389. I would like to thank Dr Agnieszka Ayşen Kaim for drawing my attention to this publication.

63 Behar, 'Wojciech Bobowski', 50–52.

64 Barnette Miller, *The Palace School of Muhammad the Conqueror* (Cambridge: Harvard University Press, 1941), 197, n. 6; Magni, *Quanto di più curioso e vago ha potuto raccontare*, I: 500, 604. Even though Cahit Öztelli suggested that A.B. must have been alive in 1676, Gültekin Oransay demonstrated that Öztelli's assumption is incorrect: see Öztelli, *Uyan Padişahım* (Istanbul: Milliyet Yayınları, 1976), 88, compare Oransay, 'Santurcu Ali Ufki Beğ Konusunda Birkaç Düzeltme', *Türk Folklor Araştırmaları* 17, no. 325 (August 1976): 7721–7723 (7722). Although Antoine Galland mentions his name in his journal on 25 January 1673, the entry is about a book he bought that 'came from Ali Beg [*qui venoit d'Ali Beg*]', which might refer to his estate as he did not write that he bought it from A.B.; see *Journal d'Antoine Galland pendant son séjour a Constantinople (1672–1673)*, ed. Charles Schefer, vol. 2 (Paris: Libraire de la Société Asiatique, 1881), 13.

Thus, A.B. most probably died during the early stages of the war between the Ottoman Empire and the Polish-Lithuanian Commonwealth (1672–1676), which was fought over the right of the Cossack Hetman Petro Doroshenko to become an Ottoman vassal and thus bring part of Ukraine under Ottoman suzerainty.[65] Dariusz Kołodziejczyk credits A.B. with 'efforts to reconcile the Polish envoy [Franciszek] Wysocki with the chief dragoman, Nicussios Panaiotti, and prevent the Polish-Ottoman war'.[66] A.B. met Wysocki in Edirne when he was the acting chief dragoman while Panaiotti was in Crete in the first half of 1670,[67] but if he indeed tried to make peace between his two homelands it did not work out. Ottoman forces conquered Kamieniec in August 1672 and besieged Lviv in September. One wonders whether he died because his heart could not bear the prospect of the invasion of his hometown by the forces of the empire he served.

A.B.'s Works and the *Janua linguarum reserata*

A.B. is known for a fairly long list of works on several topics, including music. Since his musical work contributed posthumously to the understanding of the Ottoman Turkish musical tradition by a western European audience, his writings on music and musicology might well be seen as closely related to translation and mobility, the central themes of this volume. That said, the technical aspects of his song collection, including his incomplete Turkish adaptation of the Genevan Psalter – for which he composed new Turkish lyrics and rearranged the melodies according to *makam*s (the melodic modes used in Ottoman Turkish music) – as well as the parts of his private notes and other writings that include musical notations and music-related writings, have already been studied extensively by musicologists.[68] Lacking their technical

65 Dariusz Kołodziejczyk, *Polish diplomatic relations (15th–18th century): an annotated edition of 'ahdnames and other documents* (Leiden: Brill, 2000), 146–147.

66 Kołodziejczyk, *Polish diplomatic relations*, 184.

67 François Pétis de la Croix, *Guerres des Turcs avec la Pologne, la Moscovie et la Hongrie* (The Hague: Jacques Garrel, 1689), 9.

68 Elçin, *Ali Ufkî*; Şükrü Elçin, 'Ali Ufki'nin bilinmeyen besteleri', *Türk Kültürü* 121 (November 1972): 48–51; Behar, *Ali Ufkî ve Mezmurlar*; Behar, *Saklı Mecmua*; Judith I. Haug, *Der Genfer Psalter in den Niederlanden, Deutschland, England und dem Osmanischen Reich (16.–18. Jahrhundert)* [Tübinger Beiträge zur Musikwissenschaft, vol. 30] (Tutzing: Hans Schneider, 2010); Haug, *Ottoman and European Music in 'Alī Ufuḳī's Compendium*; Judith I. Haug, *Ottoman and European Music in 'Alī Ufuḳī's Compendium, MS Turc 292: Analysis, Interpretation, Cultural Context – Volume 1: Edition* and *Volume 2: Critical Report* [Schriften

expertise, I shall not discuss the musical elements of A.B.'s work, beyond a few comments (in the third part of this chapter) on some of his original lyrics and the place of 'popular' music in his collection.

The rest of A.B.'s intellectual output, with the exception of some medical writings found among his private notes,[69] can be divided into his original writings on historical, linguistic and religious topics on the one hand, and translations on the other. Among his historical works, the first is his account of the murder of Kösem Sultan, the Queen Mother. A.B.'s original narrative, known to have been composed in Italian, does not seem to be extant, or has not yet been found. Hannah Neudecker has edited the manuscript of an English translation she found at the Bodleian Library.[70] This translation was first published among the papers of Isaac Barrow, who visited Istanbul in the late 1650s and might have been the translator of the piece, and it was also used as a source by Paul Rycaut.[71]

A.B.'s other historical work is his account of the Ottoman imperial palace and the education of the pages there, the *Saray-ı Enderun*. It might be an expanded version of his account of the palace that he shared with the Swedish ambassador Rålamb.[72] He also wrote this text in Italian, at the latest in 1665, which is the date that appears on some surviving manuscripts. A German translation was published in 1667 – the only one of his works to appear in print during his lifetime. In 1679, Magni published a redacted version written by A.B. in 1670. A French version was prepared in 1686 by the French ambassador Girardin.[73] Clearly, A.B. produced this account for the Europeans who visited Istanbul, and he probably made some money by producing individual copies for different

zur Musikwissenschaft aus Münster, Band 26, nos. 1 and 2] (Münster: Institut für Musikwissenschaft der Westfälischen Wilhelms-Universität, 2020); Cevher, *Ali Ufkî Bey ve Hâzâ Mecmû'a-i Sâz ü Söz* (Izmir [published by the author], 2003).

69 Judith I. Haug, 'Medical Knowledge in 'Alī Ufuḳī's Musical Notebook (Mid-17th Century)', *Intellectual History of the Islamicate World* 6 (2018): 117–143.

70 Hannah Neudecker, 'An Ottoman palace revolution as witnessed by a court musician', *DSNELL* 3, nos. 1–2 (1997): 163–192. The expression 'ex Italico' ('from the Italian') appears on the first page of the manuscript under the title (see 171, 191); there are also several Italian words in the margins (168–169).

71 Isaac Barrow, *The Works of the Learned Isaac Barrow, D.D., late master of Trinity-College in Cambridge*, vol. 4 (London, 1687), 196–210; Neudecker compares the text of the manuscript she edited with Rycaut's account of the same events: see Neudecker, 'An Ottoman palace revolution', 164–167.

72 Rålamb, 'A relation', 703–705.

73 There is another French translation from 1666 held at Harvard; since I was unable to consult this version, I am not sure of its relationship to the 1686 text; see Berthier and Yerasimos, *Topkapi*, 14; Fisher and Fisher, '*Topkapı Sarayı*', 7.

readers.[74] Magni even claims that A.B. copied out his account in exchange for alcohol,[75] although I suspect that he had many other means to procure alcohol in late seventeenth-century Istanbul.

In some of his works, A.B. described himself as a 'master of languages' (*magister linguarum*).[76] Not only did he teach Turkish to some students who became foundational names in the early history of Turkology and Oriental Studies in Europe, such as Meninski and Antoine Galland (d. 1715),[77] he also produced teaching materials, including a book of dialogues, written in facing columns in Turkish and French,[78] and a Turkish grammar for readers of Latin which has recently been published.[79] Yet another item one should add to his long list of works is his short selection of Turkish proverbs, *Adagia quaedam Turcica*, which includes 49 proverbs and their Italian translation.[80]

A.B.'s writings on religion consist of two Latin treatises on Islam. While neither carries a date, the *Epitome Fidei & Religionis Turcicae*, or the 'Epitome of the Turkish Faith and Religion', which is probably the older of the two, has only recently been conclusively identified by Thomas Matthew Vozar as a work by A.B. As Vozar points out, this text, which was also published among the papers of Barrow and was long believed to have been written by him,[81] must have started circulating in Europe via visitors to Istanbul in the 1650s. Rålamb's summary

---

74 Tommasino, 'Travelling East', 43, 46.

75 A.B./Magni, *Quanto di più curioso e vago ha potuto raccontare*, I: 501. A.B. seems to have had or developed later in life a drinking problem, as the Habsburg representative Giovanni Battista Casanova referred to him as a 'drunkard' in 1668; see Hering, 'Panagiotis Nikousios als Dragoman der kaiserlichen Gesandtschaft in Konstantinopel', 161, n. 98.

76 See his translation of the Church of England catechism, *Ta'lîm Ortodoksîn – Doctrina Christiana*, University of Glasgow Library, Hunter 352, the unnumbered page following 43; see also his letter to Isaac Basire in Neudecker, 'From Istanbul to London?', 191; and in his grammar, he introduces himself as a professor of Turkish: see A.B., *Grammatica Turcico-Latina*, Bodleian Library, Hyde 43, cover page.

77 Behar counts Jakab Harsányi Nagy among A.B.'s students, see 'Wojciech Bobowski', 31, n. 55.

78 BnF, ms Turc 235.

79 Ali Ufkî Bey, *Grammatica Turcico-Latina*. As is usual with A.B.'s manuscripts, the one in which this work is found also contains other things, including two Turkish songs with their lyrics that were published by Elçin in his 'Ali Ufki'nin bilinmeyen besteleri'.

80 Barrow, *The Works*, vol. 4, 192–195. A.B.'s authorship of this small Turkish proverb collection is acknowledged by John Worthington in a letter to Samuel Hartlib dated 26 October 1661: see J. Crossley and R.C. Christie, eds., *The Diary and Correspondence of Dr. John Worthington*, vol. 2, part I (Manchester, 1847–1886), 64; see also Thomas Matthew Vozar, 'Isaac Barrow, Ali Ufki and the *Epitome Fidei et Religionis Turcicae*: a seventeenth-century summary of Islam in the European Republic of Letters', *Journal of the Warburg and Courtauld Institutes* 85 (2022): 145–163 (149–150).

81 Barrow, *The Works*, vol. 4, 173–185.

of Islam in his travelogue seems to be a selective short translation of this text.[82] Vozar also mentions an English translation of the text in a manuscript from the second half of the seventeenth century that he did not have a chance to consult.[83] A modern English translation of the text has been published recently, although the editor's attribution of the Latin text to Barrow in this new translation must be corrected.[84] In terms of content, one characteristic of the text is A.B.'s sympathetic depiction of Islam and Muslims. Indeed, this caused a great deal of consternation among later scholars, because of the mistaken attribution to Barrow, who had elsewhere written quite disparagingly about Islam.[85]

A.B.'s second treatise on religion was published by Thomas Hyde in 1690 in Latin.[86] It is quite likely that he wrote it at the request of Thomas Smith, who had arrived in Istanbul as chaplain to the English ambassador Sir Daniel Harvey in 1668. It shares some content with the earlier *Epitome Fidei*, such as the section on daily prayers, pilgrimage to Mecca and circumcision; but it also includes additional elements, such as sections on judicial and religious functionaries, as well as descriptions of the practices of visiting the sick and burial of the dead. Other original elements are not strictly related to religion at all, such as a discussion of the titles of the Crimean Khan, the Ottoman emperor and the grand vizier. In 1712, the Latin text appeared in English translation in a volume alongside other works on Islam. As noted by David Pailin, there is a distinct discrepancy between 'the negative elements in Hyde's prefatory comments' regarding Islam and the tenor of A.B.'s text.[87] Furthermore, the other texts that accompany A.B.'s 'A Treatise concerning the Turkish Liturgy' share a similarly dispassionate – if not sympathetic – approach to Islam as evidenced by the two pieces by Adriaan Reland that constitute the first two books of his *De religione Mohammedica* (1705), a book that came to be known as the first objective study of Islam in Europe.[88]

---

82 Rålamb, 'A relation', 705–707; Vozar, 'Isaac Barrow', 152, n. 26, 159.

83 Vozar, 'Isaac Barrow', 158, n. 53.

84 Daniel Perett with Clinton Bennett, 'Annotated translation of *Epitome Fidei* ...', in Bennett (with Latin translation by Perett), *Promise, predicament and perplexity: Isaac Barrow (1630–1677) on Islam* (Piscataway: Gorgias Press, 2022), 25–62.

85 Vozar, 'Isaac Barrow', 151.

86 Thomas Hyde, ed., *Tractatus Alberti Bobovii* [...] *de Turcarum Liturgia, Peregrinatione Meccana, Circumcisione, Aegratorum Visitatione, &c.* (Oxford: Theatrum Sheldonianum, 1690).

87 David A. Pailin, 'British views on religion and religions in the age of William and Mary', *Method & Theory in the Study of Religion* 6 (1994): 349–375 (369); for comments on Hyde's footnotes to A.B.'s text, see Duygu Yıldırım, 'Comparing Faiths: Religious Dialogue between the Ottoman Empire and Early Modern Europe', *History of Religions* 63 (2023): 198–227 (219–222).

88 See the first two texts in *Four Treatises*; on Reland, see Lot Brouwer, 'Adriaan Reland's Leg-

As for his translations, A.B.'s earliest extant one, as mentioned above, was the translation of the catechism of the Church of England from Latin to Turkish. One of the two extant copies of the work, which A.B. dedicated to John Bendish, carries the date of 1654.[89] Although both copies ended up in England, they were not the source of a Turkish translation of a catechism published in 1660, suggesting that A.B.'s translation did not circulate much.[90] Since the original text of the catechism, first composed in 1549, was in English, Isaac Basire or Thomas Bendish must have provided A.B. with a Latin translation as, I believe, A.B. did not know English as well as it is sometimes assumed.[91] I was not able to identify the Latin text among the few Latin translations of the Anglican catechism I consulted, so it might well be Basire's own translation.

A.B.'s is a striking text in terms of its translation choices, which may reflect a particular theological commitment specific to A.B. The catechism consists of twenty-five questions and answers that a child is expected to learn by heart before confirmation, the first simply asking the child his or her name. The third question asks what the child's godparents did at his/her baptism. The child's response is supposed to be the following:

> They did promise and vow three things in my name. First, that I should renounce the devil and all his works, the pomps and vanity of this wicked world, and all the sinful lusts of the flesh. Secondly, that I should believe

acy as a Scholar of Islam', in *The Orient in Utrecht: Adriaan Reland (1676–1718), Arabist, Cartographer, Antiquarian and Scholar of Comparative Religion*, ed. Bart Jaski, Christian Lange, Anna Pytlowany and Henk J. van Rinsum (Leiden: Brill, 2021), 44–64.

89 For the extant manuscripts of this work and a general introduction to it, see Neudecker, 'Two Hitherto Little-Studied Turkish Translations'.

90 Neudecker, 'Two Hitherto Little-Studied Turkish Translations', 341; on the 1660 publication, which is a Turkish translation of John Ball's *A Short Treatise containing all the Principal Grounds of the Christian Religion* by William Seaman and appears mostly in a volume with Edward Pococke's Arabic translation of Hugo Grotius's *De veritate religionis Christianae* (Latin original published in 1640), see Falconer Madan, *Oxford Books: A Bibliography of Printed Works relating to the University and City of Oxford or Printed and Published there*, vol. 3 (Oxford; Clarendon Press, 1931), 122, no. 2480; and Noel Malcolm, 'Comenius, Boyle, Oldenburg, and the Translation of the Bible into Turkish', *Church History and Religious Culture* 87 (2007): 327–362 (340, n. 38).

91 Neudecker assumes that A.B.'s English was as good as his Latin, citing his account of the death of Kösem Sultan in 1651: see Neudecker, 'Two Hitherto Little-Studied Turkish Translations', 343 n. 45. This account, however, was originally written in Italian and translated into English, and there is no evidence to suggest that A.B. translated it himself; on the contrary, the note on the manuscript of the translation 'from the Italian' suggests a different translator: see note 70 above.

> all the articles of the Christian faith. And, thirdly, that I should keep God's holy will and commandments, and walk in the same all the days of my life.[92]

Interestingly, when translated into Latin in A.B.'s source text, the phrase 'the articles of the Christian faith' is rendered as 'Fidei articulis' rather than 'Fidei Christianæ articulis'. A.B. then translated this into Turkish as '*dîn-i İslâm'ın bâbları*', or 'the chapters of the religion of Islam'.[93]

Similarly, in the fifth question, in which the catechist asks the child to rehearse the 'articles of belief', the phrase 'fidei articulos', is translated as '*dîn-i İslâm'ın bâbları*'.[94] The same phrase comes up again in the sixth question in which the catechist asks the child what s/he learns in 'these articles of belief' or '[*b*]*u dîn-i İslâm'ın bâbları*'.[95] It is worth noting that elsewhere in the text A.B. translates the word faith (fide) rightly as '*îmân*' and faithful (fidelibus) as '*mü'min*', for example in the child's answers to the nineteenth, twenty-third, and the twenty-fifth questions.[96] Thus, it appears that he did not necessarily use Islam as a general substitute for the word faith or '*îmân*', but rather in reference to the name of 'the faith' that is defined specifically in the articles enumerated, which he translates without introducing any changes. In this way, the belief in a trinitarian understanding of God as defined in the English catechism is rendered in A.B.'s translation as 'Islam'. Whether this was something he introduced into the text on his own initiative or was rather the result of discussions with Basire is a question we can, unfortunately, not answer. Yet it definitely fits with A.B.'s ecumenical approach to religion that, as I touch upon in the next section, can be discerned in the poems he chose to copy into his collection of songs.

Perhaps A.B.'s most popular work was his Turkish translation of the Bible. While not the very first Turkish-language Bible,[97] and one that was largely

92 https://www.churchofengland.org/prayer-and-worship/worship-texts-and-resources/book-common-prayer/catechism (accessed 29 December 2023).

93 MS Hunter 352, 3; 'Articulos fidei Christianæ' is the phrase found in the 1560 Latin translation of the Anglican catechism: see William Keatinge Clay, ed., *Liturgies and Occasional Forms of Prayer set forth in the Reign of Queen Elizabeth* (Cambridge: Cambridge University Press, 1847), 417.

94 MS Hunter 352, 6; the question is formulated as 'Dic eugo fidei articulos' in the Latin source text A.B. had; compare 'Recita articulos fidei', Clay, *Liturgies*, 417.

95 MS Hunter 352, 9.

96 MS Hunter 352, 36, 40, 43.

97 Hannah Neudecker, *The Turkish Bible Translation by Yahya bin 'Ishak, also called Haki (1659)* (Leiden: Het Oosters Instituut, 1994).

ignored during his own lifetime, it was destined to become, a century and a half after his death, the most widely read of its kind. In the 1660s, when he composed the translation, there were in fact two separate European projects to produce Turkish translations of the Bible. One, based in England, was spearheaded by Robert Boyle (d. 1691), a natural philosopher and one of the founders of modern chemistry, as well as a devout Christian who longed to see the Ottoman Turks converted to Christianity. As such, Boyle also sponsored the Arabic translation of Hugo Grotius's *De veritate religionis Christianae*, which included, among other things, refutations of Islam.[98] Thanks to his sponsorship, as well as some support from the merchants of the Levant Company, the translation of the New Testament by William Seaman (d. 1680), who had earlier translated a version of John Ball's catechism which seems to have circulated alongside the Arabic translation of Grotius's *De veritate*, appeared in print in 1666.[99]

The other translation project, in which A.B. became involved, was led by Comenius, the author of the *Janua linguarum reserata*. Comenius was the last bishop of the Unity of Brethren, one of the oldest Protestant churches in Europe, founded in Bohemia in the mid-fifteenth century. The Brethren had to leave Bohemia in the early seventeenth century after the Bohemian Revolt by Protestant Bohemian noblemen against their Catholic king was crushed at the Battle of White Mountain in 1620. Under his leadership, many brethren relocated to Leszno in Poland, where for a time they were tolerated. Comenius then visited Sweden and England and eventually moved to the Netherlands as his support for the Protestant Swedes during the Second Northern War led to attacks against him in Leszno. His project to translate the Bible into Turkish developed in the Netherlands. Nicolette Mout points to Comenius's millenarian ideas as one of the factors that might have prompted him to lead a project like this. The prophesied conversion of the Ottoman Turks to Christianity was believed to be one of the expected signs of the end of times.[100]

98 Dietrich Klein, 'Hugo Grotius' Position on Islam as Described in *De veritate religionis Christianae, Liber VI*', in *Socinianism and Arminianism: Antitrinitarians, Calvinists and Cultural Exchange in Seventeenth-Century Europe*, ed. Martin Mulsow and Jan Rohls (Leiden: Brill, 2005), 149–173; G.J. Toomer, 'Edward Pococke's Arabic Translation of Grotius, *De Veritate*', *Grotiana* 33 (2012): 88–105; Charles G.D. Littleton, 'Ancient Languages and New Science: The Levant in the Intellectual Life of Robert Boyle', in *Republic of Letters and the Levant*, 151–171; and Travis Dumsday, 'Robert Boyle on the diversity of religions', *Religious Studies* 44 (2008): 315–332; and especially Malcolm, 'Comenius, Boyle, Oldenburg, and the Translation of the Bible into Turkish'.

99 Şükrü Elçin's assumption that it was A.B. who translated this text into Turkish is mistaken; see his *Ali Ufkî*, xii.

100 M.E.H.N. Mout, 'Calvinoturcismus und Chiliasmus im 17. Jahrhundert', *Pietismus und Neuzeit* 14 (1988): 72–84 (78).

Comenius secured financial support for his project from Laurens de Geer, 'the philanthropic son of a rich merchant-industrialist'.[101] Jacob Golius, who taught mathematics and eastern languages at the University of Leiden, was to supervise the translation, which was to be done by Levinus Warner, a German-born Orientalist who had been a student of Golius at Leiden University and had been in Istanbul since 1645. However, as Levinus had become the chief Dutch diplomatic officer in the Ottoman capital in 1654, and was too busy to do it himself, he passed on the commission in 1658 to his personal dragoman Haki Yahya bin Ishak, who completed his translation in 1659.[102] Warner seems to have asked A.B. his opinion of Haki's translation, as we have a gloss of his on Haki's text that is quite critical of the language.[103] Warner then asked A.B. to do a new translation, which he completed in 1664. But a Damascene Armenian, Şahin Qandi, who was in the Netherlands and was consulted by Golius, found A.B.'s text 'inadequate, boorish and barbaric', and stated that 'no educated Turk would be able to read it'.[104] The death of Levinus Warner in 1665, of de Geer in 1666 and finally of Golius in 1667 left Comenius without either a sponsor or an expert who could oversee potential amendments to the translation. Comenius himself died in 1670. Thus, A.B.'s Turkish Bible remained at the Library of Leiden University for almost 150 years without reaching its intended readership. Comenius's dedicatory preface to Mehmed IV, which he finalised in early 1666, was not published until 2007.[105]

The text received a second lease of life, however, when a large part of A.B.'s translation was eventually revised by the German orientalist Heinrich Friedrich von Diez (d. 1817) in the last years of his life. He was truly impressed by A.B.'s work, writing:

---

101 Malcolm, 'Comenius, Boyle, Oldenburg, and the Translation of the Bible into Turkish', 329.

102 Hannah Neudecker edited Haki's Turkish translation of the two books of Samuel from the Old Testament: see Neudecker, *The Turkish Bible Translation*, 13–212.

103 A.B. likens Haki's translation to a Turkish Talmud for the obscurity of its language; see the full Latin text of the gloss in Barbara Flemming, 'Two Turkish Bible Manuscripts in Leiden as Middle-Ottoman Linguistic Monuments', trans. John O'Kane, in Flemming, *Essays on Turkish Literature and History* (Leiden: Brill, 2018), 183–189 (185) (translation of 'Zwei türkische Bibelhandschriften in Leiden als mittelosmanische Sprachdenkmäler', *Wiener Zeitschrift für die Kunde des Morgenlandes* 76 [1986; *Festschrift Andreas Tietze zum 70. Geburtstag gewidmet von seinen Freunden und Schülern*, ed. Claudia Römer]: 111–118).

104 Flemming, 'Two Turkish Bible Manuscripts', 185.

105 For the original text in Latin, see Noel Malcolm, 'Comenius, the Conversion of the Turks, and the Muslim-Christian Debate on the Corruption of Scripture', *Church History and Religious Culture* 87 (2007): 477–508 (479–481); Malcolm's English translation follows, 482–485.

> If I find, in the progress of the work, Ali Bey's version as correct as hitherto, I do not say too much when I assert that it will rank among the very best versions of the sacred volume; and in many passages even excel them. His style is truly classical; .... Indeed, should the Turkish language ever be lost, it might be restored from this work in all its copiousness and ease. Having made the Turkish language for thirty years my constant study, and considered it almost a second mother tongue, it is really a treat to me to sit down in order to hear the Word of God speaking to me in this language.[106]

After von Diez's death, Jean Daniel Kieffer, another orientalist, completed the revisions. A.B.'s New Testament was finally published by the British and Foreign Bible Society in 1819 and the complete Bible in 1827. A.B.'s text continued to be revised, but it was his translation that constituted the backbone of the Turkish Bible throughout the nineteenth and early twentieth centuries, first in Arabic letters and from 1941 on in Latin letters.[107] As Agnieszka Ayşen Kaim states, his 'strategies of translation ... were very modern', which secured the longevity of his text. His language, which Şahin Qandi found to be beneath the standards of the educated elite, reflected the language as it was spoken on the streets of Istanbul. A.B. thus managed to domesticate the Bible for a vernacular-Turkish-speaking audience.[108]

Some of the expressions in A.B.'s Bible translation have been interpreted as symbols of his Christian sympathies, thus supporting his alleged wish to return to Christendom.[109] And I, too, have no doubt that when he wrote that '[i]t was love, not labour' while completing his Bible in 1664, he really meant it.[110] But we must also remember that he was 'a freelancer, who earned money through lucrative assignments that at the same time were related to his general desire for learning'.[111] While he was paid 315 *akçe*s for about 2 months at 5 *akçe*s a day as an

---

106 A.A. Cooper, *The Story of the (Osmanlı) Turkish Version with a Brief Account of Related Versions* (London: The British and Foreign Bible Society, 1901), 11, cited by Agnieszka Ayşen Kaim, 'In-between Calvinism and Islam: Ali Bey's Transcultural Translation of the Bible into Turkish in the Time of Confessionalization', *The Biblical Annals* 13 (2023): 439–461 (456).

107 For a critical edition of A.B.'s translation of the four Gospels in the New Testament, see Funda Toprak, *XVII. Yüzyıla Ait Bir İncil Tercümesi (İnceleme-Metin-Sözlük)* (Ankara: [published by the author], 2006).

108 Kaim, 'In-between Calvinism and Islam', 439, 451–454.

109 Hannah Neudecker, 'Albertus Bobovius', in *Christian-Muslim Relations: A Bibliographical History*, vol. 10: *Ottoman and Safavid Empires (1600–1700)*, ed. David Thomas and John Chesworth (Leiden: Brill, 2017), 384–403, (384–385, 397–398).

110 Neudecker, *The Turkish Bible Translation*, 372.

111 Kaim, 'In-between Calvinism and Islam', 456–457.

interpreter at the Imperial Council in 1668, when he was working on the translation of the Bible in 1664 his monthly income was between 1,800 and 2,000 *akçes*.[112] In Istanbul, working for European Christians paid much better than working for the Imperial Council.

Another translation project A.B. worked on was the above-mentioned Genevan Psalter, which was much more than a translation as it involved adapting the original to versified Turkish lyrics that could be sung according to the melody of each psalm. It is quite likely that it was also Levinus Warner who commissioned this demanding project. When Warner died in 1665, A.B. lost a patron who could finance his work and the project remained unfinished.[113] Similarly, in a letter he wrote to Isaac Basire in 1666, he mentions a plan to revise his translation of the Bible with Warner, using Theodore Beza's translation of the New Testament and his annotations, which he could now not undertake because of Warner's death. He wanted to complete this revision,[114] but never got around to doing it. While I do not mean to suggest that A.B. did all his translation for material gain, I would like to emphasise that, as these two examples show, without funding he would not have undertaken all this work for 'his former religion of the reformed profession'.

The final work of A.B. that I would like to focus on here is his Turkish translation of Comenius's *Janua linguarum reserata*, introduced at the very beginning of this essay. Here it is worth emphasising that Comenius was not just a theologian of Reformed Christianity with a missionary bend to convert the Turks; he was – and still is – better known as an educator. While Comenius was alive, the *Janua linguarum reserata* went through eighty editions, followed by twenty-six more published posthumously, the last in French in 1898. This bestseller of the seventeenth century consists of one thousand Latin sentences that are introduced in one hundred chapters, starting with an introduction, continuing with the beginning of the world, the elements, various animals, the human being, professions, houses, various topics of study, and ending with death, providence and a conclusion.[115] Students going through all these topics would expand not

---

112 The figures are from Elçin *Ali Ufkî*, v, n. 19, and Behar, 'Wojciech Bobowski', 29.

113 For a reproduction of A.B.'s Turkish version of the Genevan Psalter, which includes fourteen psalms as well as the transcription of its lyrics and music, see Behar, *Ali Ufkî ve Mezmurlar*, 47–104; for a detailed study, see Haug, *Der Genfer Psalter*, 481–578; there could, however, be another copy that is complete, as Thomas Hyde states that he has 'the Psalms of David in Turkish, writ with his own Hand'; see *Four Treatises*, 106.

114 Neudecker, 'From Istanbul to London?', 187.

115 For a full count of the hundred chapter headings in an English edition of the *Janua linguarum reserata*, see Johann Amos Comenius, *The Gate of Tongues Unlocked and Opened*, trans. John Anchoran (London: Thomas Slater, 1633), 283–285.

only their knowledge of these various topics, but also their vocabulary in Latin, assuming that they would be using a side-by-side translation in their native tongue: Comenius used about eight thousand words in Latin in his one thousand sentences. It is no wonder that the book became an essential textbook in grammar schools in many European countries, with translations in Czech, Dutch, English, French, German, Greek, Hungarian, Italian, Polish, Spanish and Swedish. When Comenius met Princess Christina in Sweden, he found out that 'she was learning Latin by his method' as well.[116]

Unfortunately, we do not exactly know the context for A.B.'s translation of the *Janua linguarum reserata*. Neudecker explains that when Comenius lived in the Netherlands and visited Golius, the latter shared the news that his brother Peter van Gool, who had converted to Catholicism and was living in Aleppo as a Carmelite friar, had translated the work into Arabic and that some of his Muslim friends wanted to translate the work into Persian, Turkish and Mongolian.[117] There are some references to Aleppo in A.B.'s private notebook which, unfortunately, do not include a date;[118] so, he might have been there, perhaps on his way to or from Egypt if he indeed visited Egypt. A more likely possibility is that Warner brought a copy of the *Janua linguarum reserata* with him from the Netherlands after having quite likely heard from his mentor Golius that the work was received with interest in Aleppo. Babinski identifies one of the editing hands on the Paris manuscript of A.B.'s translation of the *Janua linguarum reserata* as Warner's.[119] A.B. might have been commissioned to translate the work or may have done it as a teaching tool for his students. He seems to have finished it in 1658.[120] Babinski sees the translation as 'the first attempt to develop a program of Turkish study that would take a beginner to advanced proficiency'.[121] Behar notes that Meninski lists A.B.'s Turkish translation of the *Janua linguarum reserata* as one of the sources for his monumental *Thesaurus* of Arabic, Persian and Turkish.[122] Thus, A.B.'s translation did not only help someone regarded as a founder of Turkish studies in Europe to improve his Turkish, but also became a source for a dictionary that many Europeans consulted for centuries to come.

---

116 John Edward Sadler, *J.A. Comenius and the Concept of Universal Education* (London: George Allen & Unwin, 1966), 268.

117 Neudecker, 'Two Hitherto Little-Studied Turkish Translations', 350.

118 Elçin, *Ali Ufkî*, v; Haug, *Ottoman and European Music in 'Alī Ufuḳī's Compendium*, 91.

119 Babinski, 'World Literature in Practice', 182–183.

120 This date is at the bottom of the last page of the dictionary that is appended to the end of the translation; see BnF ms. turc 216, f. 197[a].

121 Babinski, 'World Literature in Practice', 178.

122 Behar, 'Wojciech Bobowski', 26.

One could also imagine A.B. teaching Latin with the Turkish version of the *Janua linguarum reserata* to some of his Turkish-speaking friends, such as Hüseyin Hezarfen from the Aegean island of Kos and Ebubekir from Damascus, the first of whom borrowed from him a book in Latin in the late 1660s. While A.B. helped Hüseyin Hezarfen by translating some Roman histories for his world history in the early 1670s, Ebubekir was later to translate Joan Blaeu's *Atlas Maior* into Turkish.[123] Thus, both these Ottoman authors developed, at the very least, a familiarity with Latin while they were around A.B. and used it in their works. It is therefore possible to imagine that they might have seen and benefited from the *Zerrîn ve Küşâde Dillerin Kapusı – Türkî ve Efrencî*, or *Golden and Unlocked* [*is*] *the Gate of Languages – Turkish and Latin*, the Turkish title A.B. gave his translation, which suggests that he used one of the later editions of Comenius's work in which the adjective golden was added to the title.[124]

At certain points in the *Zerrîn ve Küşâde Dillerin Kapusı*, A.B. goes beyond simple translation. For instance, in the section on astronomy, which is really about time keeping in an annual scale, the *Janua linguarum reserata* lists the months of the year, which A.B. duly translates. Then he makes an addition after the end of the 781st sentence:

> Sene-i hicri(y)e ... sâl-i şemsiyeden onbir gün ve altı sa'at ... nâkısdur. On iki şuhûr vardur, ya'ni Muharrem, Safer, ..., Zi'l-hicce. 'Arablarun, 'Acemlerün ve Türklerün tarîhidür. (The *hijri* [lunar] year is eleven days and six hours shorter than the solar year. There are twelve months, i.e., *Muharram, Safar, ..., Dhu'l-hijja*. It is the calendar of the Arabs, Persians, and Turks.)

He also translates this addition into Latin, thus adding the lunar months used by Muslims to the contents of the original *Janua linguarum reserata*.[125]

My favorite example of A.B.'s additions to the *Janua linguarum reserata* is in the twelfth sentence of the introductory chapter that is supposed to read: 'Qui cupit, capit omnia', which was translated as 'Who hath a desire perceives all' in the English translation of the work.[126] A.B. translates this with an Arabic

123 Haug, *Ottoman and European Music in 'Alī Ufuḳī's Compendium*, 72–76.

124 See, for instance, Johann Amos Comenius, *Janua aurea reserata quatuor linguarum*, trans. Nathanaël Duëz (Leiden: Officina Elzeviriana, 1644), which brings Latin together with German, French and Italian. Since this edition was published before Warner left the Netherlands, he might well have brought it with him to Istanbul.

125 BnF, ms. turc 216, 252^{a}–251^{b}.

126 Comenius, *The Gate of Tongues Unlocked and Opened*, second ed. (London, 1633), 2.

proverb, '*man jahada wajada*', which might be rendered as 'whoever strives (or struggles), gets (or finds) [it]'. And then he adds a Turkish proverb to strengthen the meaning: '*âşıka Bağdad ırak değil*, or 'Baghdad is not far for the lover', meaning a lover would not mind going all the way to Baghdad for his/her beloved. A.B. then turns around and changes the Latin original to reflect his idiomatic translation into (Arabic and) Turkish: 'Qui diligentia usus est inverit; Babilonia amatori non est longinqua'.[127]

A.B. seems to have intended to amend the very last sentence of the *Janua linguarum reserata* as well. In the quadrilingual edition that he might have used as his source text, this sentence reads: 'Iehovae Zebaoth, sanctissimae Individuae ac benedictae Trinitati, sit laus honor & gloria, in secula seculorum, Amen',[128] which might be roughly translated as: '[t]o Jehovah Sabaoth (the Lord of Hosts), the most holy Individual and the blessed Trinity, be praise, honour and glory, forever and ever, Amen'. A.B. either did not feel comfortable with the Trinity personally or may have thought of how the text could have been received by potential Muslim readers. In any case, he struck out both 'Trinitati' in the Latin original and the corresponding word in the Turkish translation. He tried to edit the statement to the best of his abilities, but a satisfactory solution that could be accepted by both Christians and Muslims could not really be found. He inserted '*vâhid*', or 'one' in the Turkish translation after deleting '*es-sâlûs*' (Ar. *al-thālūth*, Trinity) in his much-edited draft, but ended up keeping the original in the clean(er) copy of the text which was probably given to a reader from England.[129]

A.B. and the Location of the Universal

In 1656 – two years after he completed the Turkish translation of the Anglican catechism, and two years before he finished the *Zerrîn ve Küşâde Dillerin Kapusı* – A.B. copied the *Muqaddimat al-adab* by al-Zamakhshari (d. 1144). This well-known mediaeval rationalist theologian was also a linguist and had written

127 The Paris manuscript of the *Zerrîn ve Küşâde Dillerin Kapusı* does not include all the additions and the amendment of the original Latin; see BnF ms. turc 216, 426[a]; the Glasgow manuscript, however, does; see Hunter 160, 3.

128 Comenius, *Janua aurea reserata quatuor linguarum*, 320.

129 BnF turc 216, 201[b]; compare Hunter 160, last (unnumbered) page of the text; as noted by Neudecker, Hunter 160, which is in Glasgow today, was purchased from a certain Londoner, Thompson Cook, in 1663: see 'Two hitherto Little-Studied Turkish Translations', 344. Note that Hunter 160 is not a complete clean copy as many sentences in Latin have not been copied into it.

the *Muqaddimat al-adab* for Atsız, the Turkic ruler of the Khwarezmian realm, who wanted to learn Arabic. Not unlike the *Janua linguarum reserata*, the *Muqaddimat al-adab* introduced nouns in thematic clusters, which A.B. noted in subheadings written in Latin. Many manuscripts of the *Muqaddimat al-adab* included Persian, Turkish and/or Mongolian translations – again, not unlike the *Janua linguarum reserata*. I am not suggesting that the *Muqaddima* led A.B. to seek the *Janua*, which might (or not) have been the case. What interests me is how A.B. identified himself in the colophon of his own copy of the *Muqaddima*. Although the manuscript of this work eventually found its way to France – as Galland seems to have acquired it, most probably after A.B.'s death, alongside many of his other manuscripts[130] – A.B. had clearly copied it many years before Galland's arrival in Istanbul for his own personal use. It is found in a codex which also includes other items, some bearing A.B.'s copious personal notes. Thus, when he wrote the colophon, he was not 'performing' for anyone else. He chose to identify himself there as *ʿAlī ibn ʿAbd Allāh al-maʿrūf bi'l-Afranjī*, or 'Ali, the son of Abdullah, known as the Frank' – Abdullah, the 'servant of God', being the retrospectively assumed patronymic for converts to Islam, while 'Frank' was the generic name given to European Christians from outside the Ottoman territories. He even added to the colophon the phrase 'May God forgive us, our parents and all Muslims'.[131] Thus, a year after he met Rålamb, who reported that A.B. was hoping to return to Christendom, A.B. was self-identifying as a Muslim in a text that he was writing mainly for his own use.

Based on this evidence, I would neither insist that A.B. was a devout Muslim nor deny that he continued to feel an affinity to Christianity. However, I would argue that for the person he had become after he left the Sultan's palace, Istanbul was a more universal – hence, desirable – location than London would have been. To see why, it is important to realise that the 'universality' of an educational project like that envisioned by Comenius, symbolised by the *Janua linguarum reserata*, was ultimately bounded by Christianity. While one could argue that the universal as seen from Istanbul was similarly bounded by Islam, Islam provided two advantages. First, it had a designated, albeit lower, place for non-Muslims in its universe, expressed through the legal and theological category of the 'People of the Book', which Muslims extended not only to Christians and Jews but – in practice, if not in theory – also to Zoroastrians and Hindus. Second, as recently articulated by Shahab Ahmed, the hierarchical epi-

---

130 I would like to thank Dr Paul Babinski for drawing my attention to this manuscript. Behar indicates several manuscripts held at the BnF which belonged to A.B.: see his 'Wojciech Bobowski', 54.

131 For the colophon, which also includes the copy date, see BnF ms arabe 4645, f. 578^{b}.

stemology through which mediaeval and early modern Islam practically operated had a truly universalist ecumenical strain, distinct from the proscriptive elements of Islamic legalism, which could be freely enjoyed in one's private life. Islam, according to Ahmed, was able to embrace contradictions as its followers were free to explore seemingly blasphemous artistic, mystical, philosophical and theological ideas in the privacy of learned gatherings while the legal discourse that was suspicious of such explorations reigned supreme in the public sphere.[132]

Within the domain of the universalist ecumenical epistemology that was acceptable at the private gatherings of the Ottoman educated elite, being a Christian or Muslim might not have made as much of a difference as in the public sphere where the Islamic legal discourse operated. This might well have been an important point of attraction for A.B., who identified himself as a 'Sequolitano Polaccho' in a manuscript copy of his *Saray-ı Enderun* in 1665.[133] Pawlina interprets this unusual Italian word 'sequolitano', which, if not a typo by A.B., might well be a neologism by him, to mean 'secular'.[134] While I am not sure whether or not he would have meant to describe himself as a secular Polish man in the way we would understand it today, he might very well have thought of himself as a Polish man who could identify both as a Christian and Muslim at a certain level, and as a quite worldly – hence, secular – man.

Building on this insight, in this final section I will first identify a basic problem with the frequently made assumption that A.B. continued to wish to return to Christendom in the 1660s, then emphasise the missionary aspect of the kind of universalism that is represented by Comenius's *Janua linguarum reserata* and finally return to A.B. through his more personal writings; this will enable me to articulate more clearly what I mean by the universalism of the hierarchical epistemology of Islam that could offer a degree of freedom in private life and in one's internal beliefs, which would be difficult to find in Christendom at the time.

Earlier, in the first section of this chapter, I discussed A.B.'s European contemporaries' remarks about his alleged wish to return to Christendom. To a considerable extent, such assumptions have continued to be a part of twenty-first century scholarship as well. Neudecker, for instance, reiterated the claim that A.B. wanted to move to England through her interpretation of a letter that he wrote to Basire in 1666, which ends thus in Neudecker's translation:

132 Shahab Ahmed, *What is Islam? The Importance of Being Islamic* (Princeton: Princeton University Press, 2016), 301–541.

133 Pawlina, 'Life and work of Wojciech Bobowski revisited', 35, n. 31, 39, n. 51.

134 Pawlina, 'Life and work of Wojciech Bobowski revisited', 43.

> If I can ever do something for You here, do not hesitate to tell me this in three words. Such a great interest and serviceability have You showed in my affairs, when I was admitted through Your recommendation to the Court of the excellent lord envoy Thomas Bendish. I certainly remember this benefaction, and I will always remember it. Oh that I could repay you, and thank you in the same manner.[135]

According to Neudecker,

> Bobovius' hope of being able to repay Basire by helping him in his turn ... suggests that the Ottoman scholar had hopes of obtaining a position that would allow him to do the addressee favors. Since Basire was in England at the time, such a position would also have to be in England.[136]

I am not sure why one would assume that A.B. could not do a favour to Basire from Istanbul, such as sending him a Greek book or corresponding with the representative of Transylvania, where Basire taught before he returned to London. After all, in his letter he explicitly says: 'if I can ever do something for You here [*in his partibus*, or 'in these parts' in the Latin original]'.[137]

Neudecker supports her interpretation with statements found in a document entitled 'Papers concerning Ali Bei', in which the last point suggests that Isaac Basire recommended that A.B. be invited to London: 'one word of Instruction from the King to the Lo[rd] Amb[assado]r there, may easily fetch him over'.[138] Basire might indeed have recommended inviting him to London, but this does not imply that the purpose of A.B.'s letter was to solicit such an invitation. As pointed out in the 'Papers concerning Ali Bei' document with the phrase 'to the same purpose', A.B.'s letter to Basire was much more about his Bible translation than anything else.[139] The project to publish his Turkish Bible translation, completed in 1664, had stalled after Warner's death in 1665. Perhaps some financial support from England could have motivated A.B. to engage in

135 Neudecker, 'From Istanbul to London?', 186 (and 193).

136 Neudecker, 'From Istanbul to London?', 186.

137 The Latin text is in Neudecker, 'From Istanbul to London?', 191.

138 Neudecker, 'From Istanbul to London?', 194.

139 'VI. This person, with the helpe, and incouragem[en]t of the Learned Levinus Warnerus (Late Resident for the states in C[onstantino]ple) translated the Bible into Turkish: The originall whereof was sent to Leiden, there to be printed by the good Care of Jacobus Golius the profess[o]r there of the orientall Tongues. VII. The abovesaid Albertus, to the same purpose hath lately written a Letter from C[onstantino]ple unto D[octo]r Basire', cited by Neudecker, 'From Istanbul to London?', 194.

inserting a Latin translation to the text, maybe side by side with the Turkish, and revising his translation with the help of Theodore Beza's Latin translation and annotations. He says he was going to do all this with Warner 'but fate decided otherwise'.[140] Had Basire persuaded his superiors to invite A.B. to England, A.B.'s response was likely to be what he would later say to the French diplomat Girardin when offered a position and funds in France: 'I would gladly take this decision, if I were not held back by my wife whom I cannot abandon'.[141]

In short, the evidence Neudecker puts together does not really support the claim that A.B. wished to leave Istanbul. Yet Neudecker's very liberal interpretation of 'in these parts' as a reference to England in a letter written from Istanbul suggests that the assumption we come across in seventeenth-century contemporaries of A.B. – which he himself probably encouraged in order not to be treated as an untrustworthy renegade so that he could secure commissions and employment – is still shared by twenty-first century scholars; a European Christian forced to convert to Islam could not possibly be content in early modern Istanbul.

Conversion was also in the minds of A.B.'s European Christian contemporaries. Not only was the whole project of the Bible translation motivated by dreams of converting the Ottoman Turks to Christianity, as pointed out by, for instance, Malcolm,[142] but the *Janua linguarum reserata* was also seen as a tool of conversion beyond Christian Europe. After Golius shared the news from his brother Peter (who was a Carmelite friar in Aleppo) that Peter's Muslim friends were interested in translating the *Janua linguarum reserata* into Turkish, Persian and Mongolian after seeing Peter's Arabic translation, he apparently added: 'You see, Comenius, how happily your *Janua* opens a gate to the Gentiles'. Comenius replied: 'Not unto us, O Lord, but to Thy Name be the honour'.[143]

For Comenius, his educational project of Pansophia, or the possession of universal knowledge, was not an end in itself but rather a means to reach the 'ultimate objective: the conversion of all non-Christians and the unification of all Christian sects in preparation for Christ's Second Coming'.[144] Hearing the news about Sabbatai Sevi's messianic claims and the rumours about the Ten

140 Neudecker, 'From Istanbul to London?', 192.

141 See note 53 above.

142 See Malcolm's two articles referred to in footnotes 90 and 105 above.

143 Robert Fitzgibbon Young, *Comenius in England* (London: Oxford University Press, 1932), 48–49.

144 Walter W. Woodward, *Prospero's America: John Winthrop, Jr., Alchemy, and the Creation of New England Culture, 1606–1676* (Chapel Hill: University of North Carolina Press, 2010), 58.

Tribes of Israel gathering in the desert and elsewhere, Comenius encouraged Mehmed IV 'to take charge of restoring the Jewish people' in the preface he wrote for the Turkish translation of the Bible,[145] as the return of the Jews to Palestine was seen as a precursor to the Second Coming.

Comenius's combination of education with a missionary spirit was not limited to the 'Gentiles' of the Islamic world. He developed quite a following in New England in the seventeenth century, leading to later (and unfounded) rumours that he was offered the presidency of Harvard.[146] He believed that native 'Americans could be freed from their ignorance and brought willingly to Christ'.[147]

> [U]nity, guilelessness, willing accord ... [were] ... the keys, whereby the hearts of all men may be opened unto all things; and universal assent won to the truth; likewise victory over whatever is of darkness from sheer lack of knowledge, such as prevails among those races.[148]

Comenius's *Janua linguarum reserata* was used by native American students at Harvard's 'Indian College' in the mid-1650s. Yet the experiment did not work.

> Except for a handful, Indians simply refused to go to Harvard to labor over Latin texts and Christian theology ... [The new] president of Harvard, Charles Chauncy, asked for more funds to pay higher salaries for tutors, since they 'have to deal with such nasty savages' – a good indication that Indian distrust of Harvard was merited by the contemptuous attitude of its chief administrator.[149]

This is not the place to disentangle the mess created by missionary activities when they are mixed with racist attitudes. However, the problem posed by the 'universality' of a project imbued with Christian assumptions should be obvious. In this regard, I believe the conundrum confronted by A.B. in attempting to translate the *Janua linguarum reserata*'s trinitarian understanding of the

145 Malcolm, 'Comenius, Boyle, Oldenburg, and the Translation of the Bible into Turkish', 352.

146 Robert Fitzgibbon Young, *Comenius and the Indians of New England* (London: School of Slavonic and East European Studies in the University of London, 1929), 1–5.

147 Woodward, *Prospero's America*, 59.

148 Comenius to Samuel Hartlib in 1647, cited by Young, *Comenius and the Indians of New England*, 18.

149 Morton Borden, 'To educate the natives: American Indians were not as eager to attend colonial colleges as their Protestant teachers thought they should be', *American History Illustrated* 9, no. 9 (1975): 20–27 (23).

divine without offending his potential Muslim readers encapsulated this problem and helps explain why life as an Ottoman in Istanbul could hold a strong appeal for someone like him, even though he had not chosen the new course his life took when he was violently enslaved and forced to convert – an experience shared by many high-ranking members of the Ottoman imperial administration in the late mediaeval and early modern periods.

I do not mean to downplay the fact that A.B., as he clearly states in his *Saray-ı Enderun*, had some quite unhappy memories of his life as a palace slave, was dreaming of his freedom for a long time, and was deeply grateful to his patron Galatalı Kara Mustafa Pasha for getting him out.[150] His *Mecmûʿa-ı Saz u Söz*, a collection of both vocal and instrumental songs with many additional notes, most of which are pieces of verse that he jotted down, includes several similar expressions of the same sentiment, such as a couplet declaring that the suffering at the palace is worse than the poison of a murderer.[151] He also recorded the lyrics and music of a *türkî* (literally 'Turkish', *türkü* in modern Turkish: ballad, folk song) composed by a certain Mehmed, who must have been a former page like A.B., on the occasion of 'the exit (*hurûc*) of the slave boys/pages (*gılmân*) from the palace'. It starts with the line 'we have been imprisoned for so many years'.[152] In short, A.B. did not sugar-coat the suffering endured by the pages in the palace.

After that trial by fire, however, he seems to have built a good life for himself as a free man in Istanbul. I have already mentioned his friendship with Hezarfen Hüseyin and Ebubekir of Damascus. By the early 1670s, he had become the second interpreter for the Imperial Council. In music, his name is mentioned as a well-known *santur* player among several masters of late seventeenth and early eighteenth century music, including the great composer Itri (d. 1711), in an eighteenth-century chronicle of the Crimean Khanate, suggesting that he had a very good reputation far and wide.[153] He also might

150 A.B./Brenner, *Serai Enderum*, 132–135, 77, and 96–97, respectively.

151 For three slightly different versions of the same couplet, see Cevher, 'Ali Ufkî Bey ve Hâzâ Mecmû'a-i Sâz ü Söz', 63, 75, 82.

152 Cevher, 'Ali Ufkî Bey ve Hâzâ Mecmû'a-i Sâz ü Söz', 546.

153 Seyyid Mehmed Rıza (d. 1756), while enumerating famous musicians who, composed, played, or sang for Selim Giray Khan (1631–1704), refers to several well-known masters of Ottoman classical music in the late seventeenth and early eighteenth centuries, such as Itrî, Nazîm, and Hafız Post (d. 1694); among players of instruments, he names Tanburi Ali, Kemani Hüseyin, and 'Sanṭūrī ʿAlī'. Selim Giray was appointed as the Crimean Khan for the first time in 1671. A.B. might indeed have played for him as it is quite likely that just as Selim Giray Khan, A.B., too, might have taken part in the Polish-Ottoman War of 1672–1676, at least at the very beginning of it. While Seyyid Mehmed Rıza was not a con-

have had a child, as there is a Turkish chronogram on the occasion of a birth among his private notes.[154]

In addition to his wife, A.B. might have had male beloveds as well. In his collection of songs, there is one that he composed for a certain *Gurcî* Ali, 'Ali, the Georgian', who was apparently a page of Mehmed IV and the beloved of A.B., at least at some point in his life. The lyrics, which might also have been written or arranged by A.B., are as follows:[155]

Ey letâfet gülşeninin tâze açılmış güli	* Hey, the freshly bloomed rose of the rose garden of grace,
sen bu hüsnile nice 'âkılları kıldın deli	* you turned many a wise man mad with this beauty,
tîr-i müjgânın helâk etdi nice sâhib(-i) dili	* your arrow-like eyelashes destroyed many masters of the heart,
bir kemân-ebrû civânsın şîr-i Yezdânım 'Alî	* you are a lad with bow-like eyebrows, my lion-of-God,[156] Ali.

temporary of A.B., he is believed to have been of Crimean origin and to have had local sources; Yavuz Söylemez, 'Es-Seb'ü's-Seyyār fī aḫbār-ı mülūki't-Tatar (Tenkitli metin neşri, inceleme)' (PhD diss., Ege Üniversitesi, 2016), xiii–xx, 296; Seyyid Mehmed Rıza's reference to A.B. was first noted by Ruşen Ferit, *Bestegâr-Şair Nazîm: Hayatı, eserleri hakkında tetkikat* (Istanbul: Hilâl Matbaası, 1933), 13.

154 There is a chronogram A.B. composed himself on BnF, turc 292, f. 207^{b}, which gives the date 1068 (1657–1658). While there is no child name mentioned, the six lines of verse are about God creating life, the last two of which read thus: *tecellî eyleyüb kıldın kemâl-i sun'unı izhâr, getürdün bu gedâdan dahî bir dürr-i cihân-bânı / görüb mülk-i şühûda geldüğin didim ta'rîh, bi-hamdi'llâh kim gördüm şühûden sun'-ı Yezdânı – sene-i 1068*; after adding an *alif* between *bi-hamdi* and *llâh*, which the text is missing, the numerical values of the letters on the last half-line indeed add up to 1068. A.B. often copies things he reads; thus, this chronogram could belong to someone else who just had a child. However, the very personal nature of the piece makes me think that it is about his own child; and the fact that there is no name could suggest that it was a girl as one does not come across to girl's names as much in public writings.

155 As pointed out by Haug (*Ottoman and European Music in 'Alī Ufuḳī's Compendium*, 59), a version of these lyrics (without musical notation) is to be found among A.B.'s private notes under the simple title of *murabba'*, or quatrain; see BnF turc 292, f. 143^{b}; he might have heard it somewhere and jotted down or came up with it himself. In his music collection, the lyrics appear with slight changes and a description that identifies A.B. as the composer and Mehmed IV's page Ali the Georgian as his beloved and the dedicatee of the song; see Elçin, *Ali Ufkî*, 234; Cevher, 'Ali Ufkî Bey ve Hâzâ Mecmû'a-i Sâz ü Söz', 739. I followed the lyrics in the latter. One could listen to an interpretation of this song by Golden Horn Ensemble in their album *Ali Ufki* (Sony BMG Music Entertainment, 2009).

156 Ali, who was the fourth Caliph, cousin and son-in-law of the Prophet, is known as the lion of God; so in this line, A.B. likens Ali the Georgian to Ali the fourth caliph.

In A.B.'s Ottoman milieu, emotional (and physical) affection and love directed to other men was commonplace and acceptable, and Ottoman *divan* literature – that is, the world of literature that consists mostly of poetry composed by well-educated authors for a well-educated audience – is full of poems written by men for other, mostly younger, men.[157] A.B. jotted down some of the more vulgar descriptions of such relations in his musical collection; there is also a generic, short, conventional description of a male beloved in his private notebook.[158] It is impossible to tell whether his emotional (and/or sexual) interest in younger men might have had an impact on his preference of Istanbul over London, but if he were indeed interested in same-sex relations,[159] Istanbul may well have been the better option. As it happens, there is an early seventeenth-century love story by Atayi (d. 1635) that involves two bi-national (Ottoman and English) couples, Tayyib and Sir John and Tahir and Janno, who, after living in England for a while, experience an enforced separation because English authorities found such relations unacceptable. At the end, the Englishmen become Muslim, Sir John changing his name to Mahmud and Janno to Mesʿud, and they all move to Istanbul where they end their days happily.[160]

When A.B. wrote for a European audience, however, he sounded critical of close relationships between men. In most versions of the *Saray-ı Enderun*, there is a section devoted to same-sex relations, which includes among other things the following (as well as some more critical remarks):

---

157 There is a large literature on this topic; for an introduction and further references, see Baki Tezcan, 'Male Same-Sex Relations and Gendered Patronage Practices in a Slave Society: Understanding the Social Context of Homoerotic Relations between Janissaries', in *The Janissaries: Socio-Political and Economic Actors in the Ottoman Empire (17th-Early 19th Centuries)* [Halcyon Days in Crete XII – A Symposium Held in Rethymno, 12–14 January 2024], ed. Yannis Spyropoulos (Rethymno: Crete University Press, 2025), 385–409.

158 The short description in BnF ms. turc 292, f. 57ª, was noted by Behar, *Saklı Mecmua*, 50; for examples in A.B.'s music collection see, for instance, Cevher, 'Ali Ufkî Bey ve Hâzâ Mecmûʿa-i Sâz ü Söz', 92–94; Elçin, *Ali Ufkî*, 305, 312. Kaim identifies a sexual prank A.B. jotted down in Polish: see her 'In search of a Polish layer under the Ottoman identity of Albert Bobowski Alias Ali Ufkî', 81.

159 Agnieszka Ayşen Kaim identified A.B. as bisexual: see her 'The Adventures and Affairs of Some of the Polish Émigrés in the Ottoman Empire from the 17th to the 19th Centuries', in *Remarks on Turkish Studies / Türklük Mülahazaları*, ed. Marzena Godzińska, Aslı Cevgin Uysal and Władysław Chilmon (Warsaw: Department of Turkish Studies and Inner Asian Peoples, Faculty of Oriental Studies, University of Warsaw, 2015), 48–57 (52).

160 In the story, Sir John and Janno are introduced as Europeans, yet Sir John sounds like an English name; for a more detailed summary of the story see Walter G. Andrews and Mehmet Kalpaklı, *The Age of Beloveds: Love and the Beloved in Early-Modern Ottoman and European Culture and Society* (Durham: Duke University Press, 2005), 59–62.

> Those who hold high offices are good at disguising themselves. It happens that they wait at the window of the room to see when the chosen one goes to the mosque or another place; then they send him a gift and offer their patronage and protection. Over time they share their property and position at court. This is one of the main reasons why so few renegades who have abandoned Christianity return to their native countries (even though they often have good opportunities to do so). Instead, many of them stay with their *ahiret babası* (adoptive father), who is connected to them with love, rather than going back to their real fathers.[161]

One wonders, of course, whether A.B. could be talking about himself when he refers to renegades who would not return to their fatherlands because of the emotional connection they felt toward their adoptive fathers. Girardin, while adapting A.B.'s *Saray-ı Enderun*, expands this part a bit and claims that when he offered safe returns to Christendom to some renegades, they rejected his offer:

> … most would easily return to the fold of the Church, if they were not committed to these infidels by stronger and truer bonds of friendship than they could find among us. They also tried to persuade me that they loved these friends in all possible purity and that they had for each other only a mutual desire to be of service to each other in all encounters, and to refuse each other nothing; they exaggerated the perfection of this love and told me that, to learn to love God perfectly, it was first necessary to love his creatures with an improper and imperfect love.[162]

Coincidentally, in 1669, Girardin had indeed offered A.B. his help to go to France. As mentioned above, A.B. had turned this offer down by referring to his wife.[163]

---

161 Here I mostly follow the German translation of A.B. by Brenner, *Serai Enderum*, 70–71; the part in parentheses might be his addition to the text. Magni's text does not seem to include this section, which might be his intervention or A.B.'s – impossible to know. The French version by Girardin seems to have been expanded: see Berthier and Yerasimos, *Topkapi*, 85–86; for an Italian version that includes this section, see Pawlina's Polish translation, *Bobovius – Ali Ufkî*, 123–124. The parts I edited out of the German translation include references to the devil which do not appear in Pawlina's Polish translation. The continuation of the text has some further differences between the versions that make it difficult to establish what A.B. might have written himself.

162 Berthier and Yerasimos, *Topkapi*, 86; compare A.B./Brenner, *Serai Enderum*, 71–72; Pawlina, *Bobovius – Ali Ufkî*, 124–125.

163 See note 53 above.

Another aspect of A.B.'s private life that becomes apparent in his private notes and his music collection is the attraction he seems to have felt toward certain mystical interpretations of Islam. In the private notebook, Behar identified two legal opinions, or *fetvâs*, that A.B. made copies of; both relate to mystical questions.[164] In one of them, a mufti identified as 'Balizade Mehmed' is approached with a question about what to do to people who reproach the followers of the Halveti Sufi order for their practice of *devrân*, a ritual in which Halvetis move their bodies in the company of music while uttering devotional statements involving the names of God. The mufti is supportive of the Halvetis and their ritual and states that their critics should be disciplined.[165] While I could not identify a 'Balizade Mehmed' who was a mufti, there is a Balizade Mustafa (d. 1662), who was grand mufti for six months in 1656–1657, just after A.B. left the imperial palace. As the tension between the revivalist Muslims known as Kadızadelis (after a certain preacher named Kadızade Mehmed [d. 1635]) and the followers of the late Halveti sheikh Sivasi (d. 1639) was rising in the mid-seventeenth century, Balizade Mustafa might indeed have been approached with such a demand for a legal opinion.[166] Even if A.B. had no interest in Sufism, a legal opinion about the Halveti *devrân* would be of interest to him as a musician who copied several devotional songs in his musical collection and private notes. The Halvetis were known to have had a close relationship with music.

The other legal opinion among A.B.'s private notes belongs to the famous Ottoman Grand Mufti Ebussuʿud (d. 1574). The question is about Ibn Arabi's (d. 1240) *Fusus al-hikam*, or the 'Ringstones of Wisdom'. Ebussuʿud explains that the parts of the book that are not compatible with the *shariʿa* do not represent Ibn Arabi's own writing but had been inserted into the text by a Jew. He adds that one needs to be very careful while reading the book.[167] This legal opinion

164 Behar, *Saklı Mecmua*, 50–51.

165 BnF turc 292, 207^{b}.

166 I did not have a chance to consult the legal opinions of this mufti; for a short biography and a reference to a collection of his legal opinions, see Mehmet İpşirli and Eyyüp Said Kaya, "Mustafa Efendi, Bâlîzâde," *TDV İslâm Ansiklopedisi* [*IA* hereafter] XXXI:294–295; Abdullah Ceyhan, "Gedizli Mehmed Efendi," *IA* XIII:550–551. On Kadızade Mehmed and some of his followers, see Baki Tezcan, 'The portrait of the preacher as a young man: Two autobiographical letters by Kadızade Mehmed from the early seventeenth century', in *Political Thought and Practice in the Ottoman Empire: Halcyon days in Crete IX – A symposium held in Rethymno, 9–11 January 2015*, ed. Marinos Sariyannis (Rethymno: Crete University Press, 2019), 187–249.

167 BnF turc 292, f. 236^{a}; compare Reşat Öngören, *Osmanlılarda Tasavvuf: Anadolu'da Sûfîler, Devlet ve Ulema, 16. Yüzyıl* (Istanbul: İz Yayıncılık, 2003), 390.

with anti-Semitic overtones is slightly more restrictive than Kemalpaşazade's well-known legal opinion on Ibn Arabi.[168] While we do not know how familiar A.B. was with Ibn Arabi directly, he definitely had an affinity with some ideas that are associated with him. One of the poems jotted down in his musical collection, with an attribution to a certain Muslihuddin, is a slightly different version of a well-known Ibn Arabi poem that includes these lines:

> My heart can take on any form – for gazelles a meadow, a cloister for monks,
> a temple for idols, pilgrim's Kaʿba – tablets of Torah, scrolls of the Qur'ân.
> I profess the religion of love; wherever – its camels turn, there lives my faith.[169]

Another poem copied in his collection of songs is by a certain Nesimi and represents some of the ideas of the historical Nesimi (ex. ca. 1417) who was influenced by, among others, Ibn Arabi. The piece starts with a striking line: 'In the eyes of lovers, blasphemy and Islam are the same'.[170] It could belong to A.B.'s contemporary Kul Nesimi (d. after 1674), another of whose poems A.B. copied among his private notes, as pointed out by Haug.[171] As it happens, Kul Nesimi is a Bektaşi poet who adopted his pen-name because of his admiration for the late mediaeval Nesimi. In short, A.B. seems to have appreciated Sufi ideas that revivalist Muslims in the seventeenth century were finding blasphemous. Some of his own pieces of poetry also fit into this general worldview.[172]

---

168 Ahmed Zildzic, 'Friend *and* foe: The early Ottoman reception of Ibn 'Arabī', (PhD diss., University of California, Berkeley, 2012), 136–138.

169 Muhyiddin Ibn 'Arabi, *Translator of Desires*, trans. Michael Sells (Princeton: Princeton University Press, 2021) 39, 41. For the Arabic poem in A.B.'s musical collection, see Elçin, *Ali Ufkî*, 187 and Cevher, 'Ali Ufkî Bey ve Hâzâ Mecmû'a-i Sâz ü Söz', 106; a rough translation could read as follows: 'Before today, I used to deny my friend – because my religion was not equivalent to his religion. / When he was kind to me, my heart cleared, and I did not find anything – other than itself, appearing between colors. / So my heart became receptive to every form – a monastery for monks and a breeding ground for gazelles, / and a house of idols, and the Kaaba of Taif – and the Torah, and the Gospel, and the Qur'an'.

170 '*ʿÂşık katında küfrile İslâm birdir*', Elçin, *Ali Ufkî*, 122; Cevher, 'Ali Ufkî Bey ve Hâzâ Mecmû'a-i Sâz ü Söz', 85.

171 Haug, *Ottoman and European Music in 'Alī Ufuḳī's Compendium*, 509; I could not locate the piece by Nesimi that A.B. copied in his musical collection among the poems in Kul Nesimi's poetry collection edited by Cahit Öztelli, *Onyedinci Yüzyıl Tekke Şairi Kul Nesîmî* (Ankara: Türk Etnografya, Folklor ve Turizm Demeği, 1969).

172 See, for instance, Elçin, *Ali Ufkî*, 188 and Cevher, 'Ali Ufkî Bey ve Hâzâ Mecmû'a-i Sâz ü Söz', 88.

A.B.'s interpretation of Islam fits well with the pluralist experiences of Islam among the palace folk. Paul Rycaut, using A.B. as his main source, talks about Christ-lovers (*hûb-Mesîhî*) at the palace. As Haug notes, A.B. refers to Ebubekir of Damascus and Hüseyin Hezarfen as Christ-lovers too.[173] Another category of non-conventional Muslims referred to by Rycaut are called *müsirrîn*, those who keep the secret (*sır* in Turkish, *sirr* in Arabic). Rycaut identifies them as atheists and suggests that the secret is the absolute denial of a deity. He also adds that there are many of them among the high administrators of palace origin as well as the judges. He counts Lari Mehmed, who was executed in 1665 for his beliefs, among them.[174]

It was thus not uncommon to hold unconventional beliefs about God in early modern Istanbul. As long as people kept these ideas to themselves and did not propagate them – which was Lari Mehmed's fatal mistake, as he was also employed as an imam at a mosque – there were no limits to what one could believe. The 'secret', for instance, is one of the key concepts of Sufism. Sufis are expected not to disclose the 'secret' openly. Early Sufi martyrs like al-Hallaj (ex. 922) were executed for disclosing the 'secret' according to later Sufis. I am not sure whether the 'secret' is really the absolute denial of a deity, as Rycaut puts it. A.B. might have simplified things for him. But it is likely to be related to ideas about *wahdat al-wujud*, or the 'unity of existence', that one comes across in Ibn Arabi's work quite often.[175] From that perspective, being Christian or Muslim does not make much of a difference.

Ottoman music, which A.B. clearly cared for, allowed its own opportunities for collaboration and exchanges across devotional borders in a cosmopolitan city like Istanbul. His private notebook includes a Greek and two Armenian songs, transliterated in Latin letters, as well as religious songs in English and German.[176] But his heart clearly beat for Turkish songs, and among all his works, his private collection of Turkish songs was the one that he perhaps worked on most over the years. While the impact of Ottoman music on Jewish hymns is well documented already in the sixteenth century,[177] it is difficult to tell whether there were intercommunal performances of Ottoman music in the

---

173 Rycaut, *The Present State of the Ottoman Empire*, 129; Haug, *Ottoman and European Music in 'Alī Ufuḳī's Compendium*, 72–76.

174 Rycaut, *The Present State of the Ottoman Empire*, 129–130; Marc D. Baer, *Honored by the Glory of Islam: Conversion and Conquest in Ottoman Europe* (New York: Oxford University Press, 2008), 114.

175 Ahmed, *What is Islam?* 26–31.

176 Haug, *Ottoman and European Music in 'Alī Ufuḳī's Compendium*, 405, 407, 408.

177 Andreas Tietze and Joseph Yahalom, *Ottoman Melodies Hebrew Hymns: A 16th Century Cross-Cultural Adventure* (Budapest: Akademiai Kiado, 1995).

seventeenth century. Yet music was to become one of the areas where Christians, Jews and Muslims worked together in the later centuries of Ottoman Istanbul. The diversity of Ottoman Istanbul, and the opportunities Ottoman music offered to cultivate that diversity, could not be easily matched by London or another European city.

Haug, who spent more time than anyone with A.B.'s private notes, underlines A.B.'s 'prejudice-free, undiscriminating and non-elitist curiosity'.[178] Mid-seventeenth-century Istanbul offered a more cosmopolitan space for A.B. to cultivate his distinctive curiosity than a European city could. His well-documented interest in 'popular' music, which is much more represented in his collection of songs even though, as A.B. notes himself in *Saray-ı Enderun*,[179] the music that the elite cared for focused on other genres, was matched by the principles of Ottoman elite recruitment. As he wrote in *Saray-ı Enderun*, in the Ottoman Empire 'all nations are equal; they can distinguish themselves only by merit and prudence, and [it brings nothing] to boast [about one's] noble blood. To gain nobility [here], one must prove nobility of spirit'.[180]

In short, I suggest that, compared with the Christian universalism of the *Janua linguarum reserata*, A.B.'s lived experience in Istanbul in the mid-seventeenth century allowed him to inhabit a more cosmopolitan world in which the Islamic universal let itself be articulated in alternate ways, including more private ones which knew of no religious – or sexual, one might add – boundaries. Despite having entered this world through the violent act of enslavement and having suffered unbearable limitations on his freedom during his Ottomanisation at the imperial palace, A.B. seems to have grown to appreciate the freedoms that a well-tended private life offered him, including the freedom to engage with Christianity as well as Islam. I suspect that he badly missed his homeland, but he found things worth loving in mid-seventeenth century Istanbul, a city that captivated him, as he confessed to Rutka, and ultimately made him stay.[181] He said as much in a *türkî* he composed 'for the crane and homesickness and love':

178 Haug, *Ottoman and European Music in ʿAlī Ufuḳī's Compendium*, 457. A.B. had a few more notebooks in which he copied texts; but none of them is as rich as BnF turc 292 in terms of music. (I am not counting his collection of songs as a notebook in this instance.).

179 A.B./Brenner, *Serai Enderum*, 76–77; compare A.B./Magni, *Quanto di più curioso e vago ha potuto raccontare*, I: 551–552; also Behar, *Saklı Mecmua*, 62–69.

180 Pawlina, 'Life and work of Wojciech Bobowski revisited', 35; compare A.B./Magni, *Quanto di più curioso e vago ha potuto raccontare*, I: 599; A.B./Brenner, *Serai Enderum*, 141; Pawlina, *Bobovius – Ali Ufkî*, 177.

181 See note 56 above.

Oh Crane, to those who ask about us in our homeland,
say that he is now a slave to a beauty.[182]

Acknowledgements

This chapter grew out of a paper I presented at *Renaissance in the Borderlands: Cultures of Humanism in the Polish and Ottoman Empires*, a conference that was organised by Giancarlo Casale and Valentina Lepri at the University of Warsaw in December 2014, and then at *Between the World and the International: Thinking with Ottoman and Islamic Pasts*, which was convened by Samera Esmeir at UC Berkeley in November 2016. I returned to this paper in 2021 and presented parts of my ongoing work, first, at the conference *Convivencia in the Mediterranean: Early modern perceptions, practices, and limits in dealing with the religious other*, organised by Sina Rauschenbach and Susanne Härtel at the Selma Stern Center for Jewish Studies Berlin-Brandenburg, in November 2021, second, in the panel 'Being an Ottoman in the Early Modern Era', which was organised by Cihan Yüksel at the Middle East Studies Association Annual Meeting in December 2022, and finally in the panel 'The Literary Mediterranean', organised by Reem Taha at the Annual Convention of the Modern Language Association in San Francisco in January 2023. I am grateful to the organisers of the 2014, 2016 and 2021 conferences and the 2022 and 2023 panels, as well as the participants and audiences at these events, for giving me opportunities to present my work and for helping me develop it further with their questions and comments. I owe additional thanks to Giancarlo Casale for inviting me to contribute to this volume, meticulously editing the first draft of the chapter, and enriching it with the questions he raised. Additionally, I would like to thank Ann Thomson and the anonymous reviewers for their comments and questions that helped me give the chapter its final form. Last, but not least, I owe debts of gratitude to Amanda Batarseh and Leonardo Giorgetti for helping me with the transcription and translation of a short text in Italian (see note 47 above), to Agata Pawlina, who answered several questions over e-mail and shared her (at the time) unpublished article with me, to Paul Babinski, who helped me by sharing his insights over e-mail (see notes 13 and 130 above), to Agnieszka Ayşen Kaim, who drew my attention to an important primary source and answered some of my questions over e-mail (see note 62 above), to Cem Behar, who provided

182 '*Türkî berâ-yı turna ve gurbet ve muhabbet*: *Turna bizim yerde bizi sorana, şimdi bir güzele kuldur diyesin*', Elçin, *Ali Ufkî*, 78; Cevher, 'Ali Ufkî Bey ve Hâzâ Mecmû'a-i Sâz ü Söz', 322.

very useful comments on my penultimate draft, to Judith I. Haug, who saved me from some embarrassing mistakes by suggesting some corrections to my penultimate draft, to my colleague Daniel Stolzenberg who kindly checked two Latin translations (see note 56 above), and to Leyla Neyzi, who spent hours at the University of Glasgow Library to photograph the manuscript Hunter 160. Dariusz Kołodziejczyk, who is one of the kindest and hardest-working colleagues I ever met, was a participant at *Renaissance in the Borderlands*. This chapter is a tribute to his scholarly productivity and the kindness he showed to me over the decades.

CHAPTER 6

The Ottomanisation of Renaissance Medicine: Hayâtîzâde's *Healing Treatises for Formidable Diseases*

Tunahan Durmaz

Introduction

This is a discussion of a late seventeenth-century medical book, *Healing Treatises for Formidable Diseases* (*Resâilü'l-müşfiyye li'l-emrâzi'l-müşkile*), written in Ottoman Turkish by Hayâtîzâde Mustafa Feyzî Efendi (d. 1692), who served as Chief Physician of the Ottoman court. This book, widely known as *The Quintet of Hayâtîzâde* (*Hamse-i Hayâtîzâde*), comprises five treatises, each one dedicated to a discussion of a specific disease: hypochondriac disease, melancholy hypochondria, syphilis, plica disease and pestilential fever.[1] Hayâtîzâde's work is noteworthy for its critical approach and methodology because it offers a comprehensive discussion of these diseases, engaging closely with the contemporary medical scholarship in Latin, which I have chosen to label for the purposes of this essay as Renaissance medicine.[2] As such, it represents a significant moment in the history of early modern Ottoman medical scholarship.

Abdülhak Adnan Adıvar, the first historian to provide a critical analysis of *The Quintet*, wrote in the late 1930s that Hayâtîzâde, because he misspells the names of the European doctors he cites, probably produced this text based

1 From here on: *The Quintet*.

2 Hayâtîzâde refers to the European physicians he cites as 'Latinate physicians' (*etibbâ-yı Latin*). As I show below, these were physicians active between the late fifteenth and early seventeenth centuries and mostly associated with the Padua Medical School. Despite recent scholarly criticisms of the term Renaissance, the term 'Renaissance medicine' has been adopted here for reasons that go beyond simple considerations of chronology and geography. Several of the cited physicians, such as Antonio Benivieni and Girolamo Fracastoro, have been traditionally associated with the medical Renaissance, and an attentiveness to their key concerns related to balancing tradition with novelty (e.g., the New World, the emergence of new diseases and the canonisation of new remedies) allow us to critically engage with Hayâtîzâde's own scholarly agenda. For a recent evaluation of Renaissance medicine, see the Introduction in Vivian Nutton, *Renaissance Medicine: A Short History of European Medicine in the Sixteenth Century* (Abingdon: Routledge, 2022).

 | DOI:10.1163/9789004731806_008

on what he gathered indirectly from foreign doctors living in Istanbul at that time.[3] He added that the book at best indicates that Ottoman intellectual circles had by then become aware of new developments in European medicine, but without indicating a revolutionary European scholarly influence.[4] In a 1963 article, Süheyl Ünver also focused on *The Quintet*. However, his analysis was merely descriptive, rather than exploring Hayâtîzâde's possible intellectual sources.[5] Therefore, Adıvar's comments, formed by the positivist ideals of Turkey's Early Republican Era, remained for decades the sole historical analysis of *The Quintet*. Interestingly enough, Hayâtîzâde and *The Quintet* were also entirely absent from the most comprehensive study of early modern Ottoman medicine, published in 2009.[6] Hence, Cengiz Şişman wrote in 2016 that the intellectual channels that may have influenced both Hayâtîzâde and his *Quintet* had yet to be explored.[7] In the 2010s, some individual chapters of the *Quintet* began to attract the attention of scholars from various fields.[8] Even more recently, a number of studies, mostly in Turkish and including several master's theses, have either focused on the linguistic features of the various copies of *The Quintet* or provided transcriptions of particular treatises.[9] Nonetheless, the

3 Abdülhak Adnan Adıvar, *Osmanlı Türklerinde İlim*, 4. Baskı (İstanbul: Remzi Kitabevi, 1982), 134.

4 For Abdülhak Adnan Adıvar's argument on Hayâtîzâde, see Adıvar, 132–135.

5 Süheyl Ünver, 'Resailü'l-Müşfiye: Hayatizade Mustafa Feyzi Hamsesi', *Dirim* 5–6 (1963): 228–231.

6 Miri Shefer-Mossensohn, *Ottoman Medicine: Healing and Medical Institutions, 1500–1700* (Albany, NY: State University of New York Press, 2009).

7 Cengiz Şişman, 'Hayatizade Mustafa Efendi (Moshe Ben Rafael Abravanel), the Physician-in-Chief of the Ottoman Palace: Marrano Legacy, Ottoman Medicine and the Sabbatean Movement', in *Transcending Diaspora: Studies on Sabbateanism and Dönmes*, ed. Cengiz Şişman (Istanbul: Libra, 2016), 84.

8 In his 2012 master's thesis on history of syphilis in the Ottoman Empire, Fatih Parlak provided a comparative analysis of *The Treatise on Syphilis* with its sources in Latin: Ömer Fatih Parlak, 'On Syphilis in the Ottoman Empire and Turkish History Writing' (MA diss., Eastern Mediterranean University, 2012). In 2014, Emre Karacaoğlu published an article on *The Treatise on Melancholy Hypochondria*, which mainly focused on the medical aspects of the treatise. Emre Karacaoğlu, 'Hayatizade Mustafa Feyzi'nin (ö. 1103/1692) Risale-i Sevda-yı Merakiyyesi', *Türk Tıp Etiği ve Tıp Hukuku ve Tıp Araştırmaları Araştırmaları Yıllığı* 7, 8, 9 (2014–2016).

9 Zuhal Kültüral and Aylin Koç, 'Ruhsal Hastalıklara Dair Bir Risâle Sevdâ-yı Merâkiyye', *Türk Kültürü İncelemeleri Dergisi*, no. 30 (2014); Hamdi Üngör, 'Hayatîzâde Mustafa Feyzî'nin Risale-i Maraz-ı Efrenc Adlı Eseri Vr. (50^b–102^b) (İnceleme-Metin-Dizin-Tıpkıbasım)' (MA diss., Marmara Üniversitesi, 2016); Ali İrfan Karnak, 'Hayatîzâde Mustafa Feyzî'nin Risale-i Humma-yı Redîe Adlı Risalesi (İnceleme-Metin-Dizin-Tıpkıbasım)' (MA diss., Marmara Üniversitesi, 2016); Aylin Koç, 'Hayâtîzâde Mustafa Feyzi'nin Plika Hastalığı Risalesi ve Eserdeki Türkçe Terimler', in *XIV. Uluslararası Büyük Türk Dili Kurultayı Bildirileri Kitabı*, ed. Rasim Özyürek et al. (XIV. Uluslararası Büyük Türk Dili Kurultayı, Bakü, Azerbaycan: Türkiye

systematic study of the complete text of *The Quintet* by historians of medicine is a project still in its beginning phases.

In his 2012 dissertation, Harun Küçük included a brief section on Hayâtîzâde within his discussions on the waning influence of Avicenna in the Ottoman world in favour of what he called 'medical humanism', a development ushered in via translations produced by converso physicians.[10] Küçük then proposed a critical re-contextualisation of *The Quintet* in his 2019 monograph *Science without Leisure*, in which he devoted a section to placing Hayâtîzâde's intellectual legacy within the broader history of 'learned' medicine in the Ottoman capital.[11] In her 2021 dissertation, Duygu Yıldırım also devoted an entire chapter to *The Quintet*, developing her hypothesis that Hayâtîzâde's identity as a Jewish convert informed his translation and enabled him to challenge traditional Islamic understandings of body and soul.[12] Most recently, Sara Nur Yıldız, in a 2023 article, has offered the most comprehensive analysis of *The Quintet* to date. By systematically comparing *The Treatise on Plica Disease* with its original sources in Latin, Yıldız has contextualised plica within the contemporary disease landscape of the Ottoman Empire and highlighted the respects in which Hayâtîzâde should be considered a trans-imperial agent involved in a project of broader cultural translation and circulation of knowledge.[13]

As a whole, this body of recent historical scholarship has underscored the historical significance of *The Quintet* and its contributions to Ottoman medicine, in a way that has focused primarily on questions of 'novelty', 'new medicine' and the cross-cultural translation of medical knowledge.[14] While also

---

Azerbaycan Dostluk İşbirliği ve Dayanışma Vakfı, Yunus Emre Enstitüsü, Yurtdışı Türkler ve Akraba Topluluklar Başkanlığı, 2019); Elif Nur Önder, 'Hayatizade'nin İllet-i Merakiyye ve Sevda-yı Merakiyye Risalelerinin Tıptaki Ahlat-ı Erbaa Düşüncesine Göre İncelenmesi ve Transkripsiyonlu Metni' (MA diss., İstanbul, Fatih Sultan Mehmet Vakıf Üniversitesi, 2020); İsmail Dönmez, 'Hayatîzâde Mustafa Feyzî'nin Risâle-i İllet-i Merâkıyye Adlı Eseri (Vr. 1[b]–32[a]) (İnceleme-Metin-Dizin-Tıpkıbasım)' (MA diss., İstanbul, Marmara Üniversitesi, 2020).

10 Bekir Harun Küçük, 'Early Enlightenment in Istanbul' (PhD diss., University of California, San Diego, 2012), 116–117.

11 Harun Küçük, *Science without Leisure: Practical Naturalism in Istanbul, 1660–1732* (Pittsburgh: University of Pittsburgh Press, 2019), 149–152.

12 Duygu Yıldırım, 'The Age of the Perplexed: Translating Nature and Bodies Between the Ottoman Empire and Europe, 1650–1730' (PhD diss., Stanford University, 2021), 173–209.

13 Sara Nur Yıldız, 'A Strange Affliction from Abroad: The Ottoman Chief Imperial Physician Ḥayātīzāde's Treatise on the Polish Plait (Plica Polonica)', *Intellectual History of the Islamicate World* 11, no. 1 (2022): 12–20, 20–27.

14 For general overviews of the Ottoman encounter with European medical knowledge,

touching to some extent upon the question of 'new', my analysis here will mainly contribute to discussions on cultural translation and the production of knowledge, as I explore how Hayâtîzâde approached certain ideas used in the contemporary medical scholarship in Latin and productively adapted them to the Ottoman cultural context. Building upon and problematising Yıldız's insight that Hayâtîzâde's translation project 'rendered Turkish a vernacular recipient of Latin language',[15] I argue that *The Quintet* is a multi-layered text whose agenda extends beyond that of linguistic translation; in other words, more than a direct semantic translation of a body of knowledge from Latin into Turkish, it constitutes a more ambitious project of cultural translation that required familiarity with both European and local knowledge. In this sense, what Hayâtîzâde conducted was a noteworthy example of medical vernacularisation.[16] More specifically, *The Quintet* is an Ottomanisation of Renaissance medicine in its knowledge and practices, achieved through the versatile (re)adaptation of medical ideas and perspectives taken from one cultural context and applied to another.[17] In that sense, it also represents a moment of encounter between two only partially commensurate epistemic cultures, out of which either novel or revised frameworks appear.

Key to this process of Ottomanisation was Hayâtîzâde's incorporation of empirical methods through observation, experience and experimentation. As he notes at the beginning of *The Quintet*, Hayâtîzâde 'scrutinised and consul-

---

see also Feza Günergun, 'Ottoman Encounters with European Science: Sixteenth- and Seventeenth-Century Translations into Turkish', in *Cultural Translation in Early Modern Europe*, ed. Peter Burke and R. Po-chia Hsia (Cambridge: Cambridge University Press, 2007), 192–211; Ebru Boyar, 'Medicine in Practice: European Influences on the Ottoman Medical Habitat', *Turkish Historical Review* 9, no. 3 (2018): 213–241.

15 Yıldız, 'A Strange Affliction from Abroad', 8.

16 On vernacular authors' refashioning of knowledge from learned medical texts in Latin, see Mary Elizabeth Fissell, *Vernacular Bodies: The Politics of Reproduction in Early Modern England* (Oxford, New York: Oxford University Press, 2004), 6. Following Fissell, I use vernacularisation here to describe the continuous authorial involvement in the refashioning of knowledge as well as the tools implemented in doing so.

17 The issue of translation and transmission of knowledge between texts, cultures and geographies has been a lively topic of discussion for scholars. See, for example, the introduction to a special issue of *Osiris*: Tara Alberts, Sietske Fransen and Elaine Leong, 'Translating Medicine, ca. 800–1900: Articulations and Disarticulations', *Osiris* 37 (2022): 1–23. For an approach focused on the influence of translation beyond textual (re)production, see Projit Bihari Mukharji, 'Casting Blood Circulations: Translatability and Braiding Sciences in Colonial Bengal', *Osiris* 37 (2022): 192–210. See also, Ronit Yoeli-Tlalim, *ReOrienting Histories of Medicine: Encounters along the Silk Roads* (London, New York, Oxford, New Delhi, Sydney: Bloomsbury Academic, 2021), 1–24.

ted' books in Latin as well as the books of ancient scholars.[18] Added to his multilingual skills of inquiry and learning was also his empiricism: he integrated his clinical observations about patients, the qualities of samples collected from them and the symptoms of diseases as seen in different patients. His citation choices also reflected this sensitivity, as he mainly cites those authors who considered case histories of individual patients, to which he added his own personal experience with certain medical conditions or methods of treatment. He also emphasised the importance of occasionally using experimental treatments. Through this combination of empirical methods, his writings go beyond decontextualised theoretical discussions, by offering medical knowledge that could meet various real-life needs for his readers.

In what follows, I offer a four-fold analysis of *The Quintet*. First, I provide a biographical account of Hayâtîzâde as a prominent figure in late seventeenth-century Ottoman courtly circles, as well as a consideration of his legacy, the posthumous popularity of *The Quintet* in the decades after its author's death, and the work's place in Ottoman manuscript culture. Second, I provide a critical engagement with recent scholarship on the question of 'new medicine', which enables me to re-evaluate the textual dynamics and objectives of *The Quintet*. Third, I explore the practical modes of writing used in *the Quintet* and question to what extent they should be considered novel methods in Ottoman medical writing. Last, I investigate Hayâtîzâde's search for new sources of knowledge, how he approached the Latin medical works, how he incorporated empirical methods into *The Quintet* and how all of this resulted in a practice of vernacularisation.

18 'bu fakir-i pür-taksir yanımda mevcûd olan kütb ve resâil-i kudemâ ve nesh-i telifât-ı Latin etibbâsının füzelâsından mütâlaa ve müzâkerem ...' ('the humble and flawed I, scrutinising and consulting the books and treatises of the ancient scholars, as well as the copies of the works [written] by the distinguished Latinate physicians'), Hayâtîzâde Mustafa Feyzi Efendi, 'Resâilü'l-Müşfiyye Li'l-Emrâzi'l-Müşkile', 1693, Süleymaniye Kütüphanesi, Selim Ağa Koleksiyonu 868, fol. 1^{b}. This quote not only demonstrates that he read the books in Latin but also suggests his abilities to read other languages, depending on the language through which he was reading the books of the ancient scholars, that is Hippocrates and Galen. Thus, Cengiz Şişman writes that he knew Arabic, Persian, Spanish, Greek, Latin, and most likely some Hebrew. See Cengiz Şişman, 'Abravanel, Moses Ben Raphael' in *Encyclopedia of Jews in the Islamic World*, ed. Norman A. Stillman and Phillip Isaac Ackerman-Lieberman (Leiden, Boston: Brill, 2010).

'Praised and Disgraced': Hayâtîzâde Mustafa Feyzî Efendi (d. 1692)

> … Luqman of the universe[s], Hippocrates of [our] time, Galen of [this] era …[19]

An unknown copyist wrote this in one of the earliest copies of *The Quintet* (c. 1689), now kept in the Hafid Efendi Collection at Süleymaniye Manuscript Library in Istanbul. His biographer, Şeyhî Mehmed Efendi confirms this adulatory view, adding that 'ilminde ferîd, fenninde vahîd idi' ('[He] was unique in terms of his knowledge and one of a kind with regard to his technique').[20] But despite this posthumous reputation, which reflects the enormous stature that Hayâtîzâde had achieved by the end of his life, his path to becoming a fully integrated member of the upper echelons of Ottoman intellectual life was an idiosyncratic one, after beginning as an outsider, with no shortage of twists and turns along the way. Since elements of his life story have been frequently noted in interpreting his work, it is important to begin our discussion by sketching out the main turning points of his career, as well as the blank spaces that still remain to be filled in.

Hayâtîzâde Mustafa Feyzî Efendi, as he would come to be known only later in life, was born Moshe ben Rafael Abravanel, a Jew from the Marrano branch of the Abravanel family, which settled in Istanbul sometime after their expulsion from Spain in 1492.[21] His date of birth is not known, nor is it clear at what point he converted to Islam, nor how he came to be recruited to work as a physician in the Ottoman palace. Cengiz Şişman highlights two different accounts of his recruitment and conversion. According to one, he converted to Islam when he was recruited in the Palace by Hadice Turhan Sultan, the powerful Queen Mother of Sultan Mehmed IV.[22] The second account, however, suggests that he

---

19 'Bu beş risale Sultan Mehmed merhumun hekimbaşısı olan lokman-ı cihan, bokrat-ı zaman, calinus-ı devran, Feyzi Mustafa Efendi Hazretlerinin telifidir …' ('These five treatises belong to Luqman of the universe[s], Hippocrates of [our] time, Galen of [this] era, that is the Exalted Feyzi Mustafa Efendi, the Chief Physician of the late Sultan Mehmed …'), Hayatizade Mustafa Feyzi Efendi, 'Resâilü'l-Müşfiyye Li'l-Emrâzi'l-Müşkile', 1689, fol. 11[a], Süleymaniye Kütüphanesi, Hafid Efendi Koleksiyonu 265.

20 Şeyhî Mehmed Efendi, *Vekayiü'l-Fuzela Şeyhî'nin Şakâ'ik Zeyli III*, ed. Derya Örs, trans. Ramazan Ekinci (Istanbul: Türkiye Yazma Eserler Kurumu Başkanlığı, 2018), 1936.

21 The most illuminating study of Hayâtîzâde's family origins is Cengiz Şişman, 'Abravanel Family', in *Encyclopedia of Jews in the Islamic World* (Leiden: Brill, 2010). Şişman also provides the most comprehensive analysis of Hayatizâde's career in Şişman, 'Hayatizade Mustafa Efendi (Moshe Ben Rafael Abravanel), the Physician-in-Chief of the Ottoman Palace: Marrano Legacy, Ottoman Medicine and the Sabbatean Movement'.

22 This story is from the chronicle of Silahtar Mehmet Ağa (*Silahtar Tarihi*). Şişman, 'Hayatizade Mustafa Efendi (Moshe Ben Rafael Abravanel)', 78.

converted sometime after his recruitment to the Ottoman Court.[23] Cengiz Şişman is convinced that he must have been recruited in the Palace in the early 1650s[24] and converted to Islam before the 1660s.[25] To his analysis, we can add that a deed concerning a property sale in Bursa, recorded in the registers of the Rumeli Sadareti Court in September 1657, suggests that he had not yet converted by that time, as the register refers to him as '… Hayâtîzâde nâm Yahudi tabib' ('a Jewish physician named Hayâtîzâde').[26]

In any case, it is clear that Hayâtîzâde was a brilliant physician with a sophisticated background that differentiated him from many other physicians of his time. As such, he attracted the attention of the palace elite, particularly the Queen Mother Valide Turhan Sultan, and was recruited in the Palace, initially as a member of either the Royal Physicians Corps (*Etibbâ-i Hassa*) or Jewish Physicians Corps (*Etibbâ-i Yahûdiyân*).[27] He also worked as a madrasa professor (*müderris*) in Süleymaniye Medical School (*Süleymaniye Tıp Medresesi*), the only medical school in the Ottoman Empire at that time.[28] In 1666, he served as translator for Sabbatai Sevi, a Jewish 'messiah' accused of sedition, during his trial before the sultan.[29] This incident gained him a significant reputation, as he was applauded as the person who convinced Sabbatai Sevi to adopt Islam and thus avoid condemnation and a potential religious upheaval.[30] Then, following the death in 1669 of Salih bin Nasrullah, the empire's Chief Physician (*hekimbaşı*), he was appointed as his successor and stayed in this position until his dismissal in 1691.[31] Considering the frequent changes of Chief Physicians in the previous periods, Hayâtîzâde's occupation of this post for twenty-two years without interruption constitutes an unusual case and attests to his success. In spite of this brilliant and long career, however, his dismissal from the post was acrimonious and presumably caused by some political and religious factionalism in the Palace.

---

23 This account is provided in the *Treatise of Kürt Hatip* (*Risâle-i Kürd Hatip*). Şişman, 79.

24 Şişman, 57.

25 Şişman, 79.

26 As noted in 'Rumeli Sadâreti Mahkemesi 106 Numaralı Sicil (H. 1067–1069 / M. 1656–1658) Hüküm No: 488' (İstanbul Kadı Sicilleri, n.d.), http://www.kadisicilleri.org/.

27 For more on his initial recruitment in the Palace, see Şişman, 79.

28 Nuran Yıldırım, 'Hayâtîzâde Mustafa Feyzî', in *Türk Diyanet Vakfı İslam Ansiklopedisi*, n.d., https://islamansiklopedisi.org.tr/hayatizade-mustafa-feyzi. For more on Süleymaniye Medical School, see Nil Sarı, 'Süleymaniye Tıp Medresesi', in *Osmanlı medreseleri: eğitim, yönetim ve finans*, ed. Şükran Fazlıoğlu and Fuat Aydın, 1. baskı, Mahya yayıncılık 80 (İstanbul: Mahya, 2019).

29 Şişman, 58.

30 Şişman, 58–59.

31 Yıldırım, 'Hayâtîzâde Mustafa Feyzî'.

Şişman provides two different accounts of Hayâtîzâde's removal from the office of Chief Physician. According to the first, he was accused of causing the death of Sultan Süleyman II, having applied the wrong treatment. This ended in his removal from the position. But there is a second version of events narrated by the chronicler Silahtâr. According to him, Hayâtîzâde was disliked by certain groups among the scholars (*ulemâs*): primarily those who were under the influence of Kadızâdeli, a religious fundamentalist movement. Under the pretext that he had discontinued public prayers, rumours began to circulate that he had converted back to Judaism. The issue was finally brought to the attention of Sultan Ahmed II, who did not initially take the accusations seriously. Unfortunately, not being able to resist the religious scholars' insistence, the sultan eventually succumbed to their pressure and had to dismiss Hayâtîzâde.[32] Hayâtîzâde died a year later, in 1692.[33] Despite this unfortunate end, his descendants would in time reach prominent posts in the Ottoman Empire. Like him, his two grandsons Hayâtîzâde Mustafa Feyzî and Mehmed Emin Efendi would both become Chief Physicians (*hekimbaşı*). While serving in this post, Mehmed Emin Efendi was also appointed to the post of Shaikh-al Islam, the highest Islamic religious authority in the Ottoman empire.[34]

Vernacular Language and Audience

Hayâtîzâde's main authored text, *The Quintet*, was commissioned by Sultan Mehmed IV sometime in the late 1670s or early 1680s.[35] It was the latter's special request that the text be produced in Turkish for the public to be able to benefit from 'the investigation and the challenge that eventually helped to under-

32 Şişman, 86–87.

33 Coming from a prominent family as well as having served as the Chief Physician for 22 years, Hayâtîzâde possessed a huge amount of wealth and left a vast heritage to his inheritors. His property and money were supposed to form a *waqf*, that is a pious foundation. However, his endowed properties could not be verified for some reason and thus the actual endowment became possible only 79 years after his death in 1771. The legal record that notes the verification of the foundation deeds as well as the list of inherited possessions was kept in the Istanbul Court: İstanbul Mahkemesi 33 Numaralı Sicil (H. 1183–1185 / M. 1769–1770), Hüküm No: 153 (İstanbul Kadı Sicilleri, n.d.), http://www.kadisicilleri.org/madde.php?klme=hayatizade&trch=hkm&-find=+ARA+.

34 Yıldırım, 'Hayâtîzâde Mustafa Feyzî'.

35 Harun Küçük notes the date of compilation as 1674: Küçük, *Science without Leisure*, 149. However, he does not provide a reference for this information. The copies I have seen go as far back as only 1686.

stand and cure these diseases'.[36] The institution of patronage, particularly royal patronage, remained the locomotive of scholarly production in early modern Ottoman intellectual circles.[37] Even though the language of learning (especially madrasa learning) was Arabic throughout this period, it had already by the seventeenth century become a common practice in palace circles and among the high-bureaucratic elite to commission scholarly books in vernacular Turkish. As Ahmet Tunç Şen discusses, the emergence of Ottoman Turkish as a scholarly language started as early as the fifteenth century but never replaced the use of Arabic as the main language of scholarly writing.[38] In the late seventeenth century, the patronage of scholarly works in Turkish considerably increased, especially in the field of medicine. Harun Küçük has observed that this period was one of medical vernacularisation.[39]

Written to accommodate similar demands, *The Quintet* was widely circulated among different audiences in the Empire and had a long afterlife as a work of medical reference. The text received such widespread attention that it became one of the most copied medical manuscripts from the late seventeenth century, so much so that Şeyhî Mehmed Efendi (d. 1731) wrote several decades later that the treatises on hypochondriac disease and melancholy hypochondria were still 'much in demand' (*mütedavil-i mergûb*).[40] The book records in the probate inventories of some eighteenth-century dignitaries also attest that the compilation had become an indispensable part of personal libraries in this

36 It is known that Mehmed IV showed special interest in European knowledge as he commissioned multiple translations of major European scholarship. A significant example is a rendering of Blaeu's *Atlas Maior* (1665) by Abû Bakr al-Dimashqî (d. 1691). For more on this translation and its early copies, see Sonja Brentjes, 'On Two Manuscripts by Abū Bakr b. Bahrām al-Dimashq (d. 1102/1691) Related to W. and J. Blaeu's Atlas Maior', *Osmanlı Araştırmaları / The Journal of Ottoman Studies* XL (2012): 171–192. In addition, Mehmed IV and his retinue showed continued interest in European medicine. Works by Fatih Çalışır and Sara Nur Yıldız have shed light on this phenomenon in recent years: Fatih Çalışır, 'An Italian Physician in the Service of the Ottomans: Giovanni Mascellini (1612–1675) and His Artis Medicae (1673)', in *Living in the Ottoman Lands: Identities, Administration, and Warfare* (Istanbul: Kronik Kitap, 2021); Yıldız, 'A Strange Affliction from Abroad', 20–27.

37 For a detailed study of the dynamics of Ottoman manuscript production see Emine Fetvacı, 'Viziers to Eunuchs: Transitions in Ottoman Manuscript Patronage, 1566–1617' (PhD. diss., Harvard University, 2005).

38 Ahmet Tunç Şen, 'The Emergence of a New Scholarly Language: The Case of Ottoman Turkish', in *Routledge Handbook on the Sciences in Islamicate Societies* (Abingdon: Routledge, 2023), 242.

39 Küçük, *Science without Leisure*, 153.

40 Meḥmed Şeyḫī Efendi, Ramazan Ekinci, and Derya Örs, *Vekâyi'u'l-fuzalâ: Şeyhi'nin Şaka'ik zeyli*, 1. baskı, Türkiye Yazma Eserler Kurumu Başkanlığı Yayınları; Bilim ve Felsefe Serisi, 93. 12 (Fatih, İstanbul: Türkiye Yazma Eserler Kurumu Yayınları, 2018), 1936.

period. For instance, his descendant Hayâtîzâde Mehmed Emin Efendi had several copies of *The Quintet* with him when he died in Damascus on his way to taking over the judgeship of Mecca.[41] The existence of at least eighty copies of the text in various libraries has been established via collection catalogues. The copies are scattered not only throughout the former lands of the Ottoman Empire but also across Europe. While most of these are copies of the complete work, some are of only a single treatise. Most are housed in various manuscript libraries in Istanbul, but some are found today in the collections of Anatolian cities such as Konya, Kütahya, Kastamonu and Samsun, among others. In addition, there are copies in the Oriental collections at the Vatican Library, the British Library, the National Library of Madrid, the German National Library, the Grand Ducal Library in Austria (Gotha Collection) and the National Library of Egypt.[42]

Such a widespread distribution raises the obvious question of audience: who read *The Quintet* and for what purposes? Finding a full answer to this question would constitute a project on its own, as it would entail a thorough examination of every available copy. Here, I will limit my analysis to the earliest copies of the manuscript. Although we only roughly know the original date of compilation of the book, an investigation conducted among the copies available in the manuscript libraries in Istanbul has revealed four copies from the 1680s and 1690s: Cerrahpaşa 473,[43] Nadir Eserler 2736,[44] Hafid Efendi 265,[45] and Selim Ağa 868.[46] A comparison of the last two provides interesting clues in relation

41 'Hayâtîzâde Mehmed Emin Efendi'nin Terekesi', H. 1160–1161 (1747–1748), fols. 50ª–53ª, İstanbul Kısmet-i Askeriyye Sicilleri, İstanbul Sicilleri. For Hayâtîzâde Mehmed Emin Efendi's biography see Mehmet İpşirli, 'Hayâtîzâde Mehmed Emin Efendi', TDV İslâm Ansiklopedisi, https://islamansiklopedisi.org.tr/mehmed-emin-efendi-hayatizade, accessed 16 July 2023.

42 This information is based on a simple search with the keyword 'Hayatizade' in the online catalogue of the General Directorate of Turkish Manuscript Libraries. In addition, Harun Küçük notes that he came across some copies written in the Armenian script: Küçük, 'Early Modern Ottoman Science': 407–408. However, I have not been able to locate them yet.

43 Hayatizade Mustafa Feyzi Efendi, 'Resâilü'l-Müşfiyye Li'l-Emrâzi'l-Müşkile', 1695, Cerrahpaşa Tıp Kütüphanesi Tıp Yazmaları Koleksiyonu 473. I have been unable to see Cerrahpaşa 473, which is dated 1695 in the catalogue and possessed by Cerrahpaşa Faculty of Medicine, because the collection was unavailable due to a removal process to another building.

44 Hayatizade Mustafa Feyzi Efendi, 'Resâilü'l-Müşfiyye Li'l-Emrâzi'l-Müşkile', 1686, İstanbul Üniversitesi Nadir Eserler Kütüphanesi, Türkçe Yazmalar Koleksiyonu 2736. This copy includes only the first treatise *Risâle-i illet-i merâkiyye* (A treatise on hypochondria).

45 Hayatizade Mustafa Feyzi Efendi, 'Resâilü'l-Müşfiyye Li'l-Emrâzi'l-Müşkile', 1689.

46 Hayatizade Mustafa Feyzi Efendi, 'Resâilü'l-Müşfiyye Li'l-Emrâzi'l-Müşkile', 1693, Süleymaniye Kütüphanesi, Selim Ağa Koleksiyonu 868.

to audience(s). On the one hand, Hafid Efendi 256 possesses the qualities of a commonplace book. Its first twelve folios are not from *The Quintet*, featuring instead *hikâyât* (stories), samples of Persian poetry and some haphasard notes about the content of *The Quintet*. The book starts somewhere after the twelfth folio. The owner, who was probably a practitioner or a medical student, includes his own notes in the margins. Since there is no seal bearing the name of a foundation, this copy was probably never a part of a library collection until much later. On the other hand, Selim Ağa 868 is arguably the cleanest manuscript of the group. All five treatises in the manuscript were copied in a clean hand, arguably by a professional scribe. Nevertheless, the front pages of this copy still appear to have been used like a scrapbook by its owner, who wrote down a broad range of notes including medical observations, drug recipes and lists of fruit and vegetable names, as well as some recipes and anecdotes from a few individuals, including the Chief Physician Nuh Efendi. This copy was therefore probably used by a practicing physician who served in the Royal Corps of Physicians at Topkapı Palace. Unless otherwise noted, my source of reference in the following empirical analysis of the reception of Renaissance medicine in *The Quintet* is from this manuscript, Selim Ağa 868.[47]

The 'New' in Ottoman Medical Writing

Throughout this chapter, and in particular in the analysis that follows, I sometimes use the adjective 'new' to acknowledge some of Hayâtîzâde's concerns that led him to compile *The Quintet*. My use requires some explanation, as I do not intend with it to imply an epistemic break.[48] This represents a critical departure from most existing scholarship, which has underscored the emergence of the so-called Ottoman 'new medicine' (*tıbb-ı cedîd*), mainly under the influence of Paracelsus, starting from the late seventeenth century.[49] It appears

---

47 The textual organisation of Selim Ağa 868 is as follows: Introduction and *Risâle-i illet-i merâkiyye* (A treatise on hypochondria), fol. 1ᵇ–32; *Risâle-i sevda-ı merâkiyye* (A treatise on melancholy hypochondria), fol. 33ᵇ–48ᵃ; *Risâle-i maraz-ı efrenci* (A treatise on syphilis), fol. 48ᵇ–95ᵇ; *Risâle-i pelika* (A treatise on plica disease), fol. 96ᵃ–111ᵃ; *Risâle-i hummâ-yı redîye* (A treatise on the pestilential fever) 112ᵇ–150ᵇ.

48 The term *epistemic break* has been used in the history of science to imply a break with 'the old' in favour of 'the new'. See Ian Buchanan, 'Epistemological Break' in *A Dictionary of Critical Theory* (Oxford: Oxford University Press, 2010).

49 This tendency is primarily articulated in the historiography of Islamic medicine: Manfred Ullmann, *Islamic Medicine*, The New Edinburgh Islamic Surveys NEIS (Edinburgh: Edin-

that, alongside the influence of European historiography, this emphasis in scholarship has been driven primarily by the increasing appearance of explicit references to the term 'new medicine' during later periods, particularly the eighteenth century, as well as certain interests of the Chief Physician Ibn Sellum (d. 1669).

Ibn Sellum's interest in novel methods is most visible with respect to chemical drug recipes.[50] This was accompanied by his interest in the writings of some European medical authorities, including, but by no means limited to, Paracelsus, the celebrated figure of Renaissance medicine who sharply rejected and even disparaged the Hippocratic-Galenic tradition in favour of a new system of 'alchemical medicine' or iatrochemistry.[51] As a result, Ibn Sellum commissioned extensive translations from contemporary medical literature in Latin. Since Hayâtîzâde mentions new diseases (*emrâz-ı cedîdân*), and the New World (*yeni dünya*) as a hub of new medicinal plants, and occasionally also refers to some chemical recipes, he has also been seen as a promoter of this shared enterprise, following in the footsteps of Ibn Sellum.[52] However, while Hayâtîzâde may have been inspired by an awareness of change and an interest in novelty, it would be an overstatement to say that he explicitly conceptualised his medical practice as a 'new medicine' self-consciously aiming to break free from Galenism. In fact, Hayâtîzâde barely implies an epistemological break in his work. To him, 'new' does not *per se* denote new medical frameworks and interests. His use of the term indicates, at best, his witnessing of certain realities of his time and, accordingly, a need for change and revision based on practical considerations. For this very reason, approaching it through an anticipation of

---

burgh University Press, 1997), 50. Harun Küçük's *Science without Leisure* represents the view that 'Paracelsian new medicine' begins in late seventeenth-century Istanbul.

50 For in-depth analyses of Ibn Sellum's scholarship, see Natalia Bachour, *Oswaldus Crollius und Daniel Sennert im frühneuzeitlichen Istanbul. Studien zur Rezeption des Paracelsismus im Werk des osmanischen Arztes Ṣāliḥ b. Naṣrullāh Ibn Sallūm al-Ḥalabī* (Herbolzheim: Centaurus Verlag & Media, 2012); Natalia Bachour, 'Iatrochemistry and Paracelsism in the Ottoman Empire in the Sixteenth and Seventeenth Centuries', *Intellectual History of the Islamicate World* 6, no. 1–2 (2018): 82–116.

51 Bachour remains hesitant about the Ibn Sellum's Paracelsism, mainly because those of his works marked as Paracelsian were only posthumously completed translations and did not necessarily follow the philosophy of Paracelsus. See, Bachour, *Oswaldus Crollius und Daniel Sennert im frühneuzeitlichen Istanbul*, 374–377.

52 Although Harun Küçük acknowledged that Hayâtîzâde was not a follower of Paracelsus, his chapter in *Science without Leisure* titled 'Recipe: An Annotated Chronology of New Medicine in the Seventeenth Century' placed Hayâtîzâde's intellectual agenda along the same trajectory of Paracelsian new medicine (*Science without Leisure*, 143–166). See alo Küçük, 'Early Enlightenment in Istanbul', 116.

specific 'new' epistemologies, particularly of a sweeping or categorical nature, may hinder our perception of what was truly at stake for Hayâtîzâde.

To reiterate, this interpretation stands in contrast to the way recent historiography has characterized Ibn Sellum, whose reception seems to have led scholars to observe something close to an epistemic break turning the new medicine into a novel ontological entity, distanced and differentiated from traditional frameworks (i.e., the humoral pathological medicine of the Galenic-Avicennan tradition). Nonetheless, in recent years, more nuanced approaches have appeared. In his 2020 dissertation, Akif Ercihan Yerlioğlu suggested that new medicine was not a static category promoting a unified medical theory.[53] Rather, 'it was a broader category of medical sciences' covering a variety of medical approaches from the late seventeenth to the nineteenth centuries.[54] Yerlioğlu's findings indicate that, even at a time when the term '*tıbb-ı cedîd*' became more frequently used, it kept eluding clear definitions as its scope also changed and expanded over time. In a subsequent 2024 article, Yerlioğlu further elaborates on this approach, as he focuses on the 1703 decree that supposedly banned new medicine in the Ottoman Empire. He not only corrects the continued misdating of the edict to the year 1704 – a mistake that has recurred in historiography since the early twentieth century – but also traced the origins of this mistake and its repercussions, such as the misperception of the decree as a total ban on Western medical practices in the Empire. Instead, he demonstrated that the ban was not on novel methods of treatment, rather on specific physicians of European origin.[55]

In her 2021 dissertation, Duygu Yıldırım likewise takes a cautious position, arguing that, rather than an adoption of the Paracelsian new medicine, new medicine 'represents a shift' from 'Islamic canon medicine' to 'medical case narrative'.[56] In a later publication, Yıldırım has reiterated her critical position towards a 'unified new' and further argued that the new served as a tool for the contemporary authors 'to cope with uncertainties' and 'the epistemic crisis of the era'.[57] Considering Hayâtîzâde's *Quintet* specifically, this latter point would

53 For more on Yerlioğlu's discussion on the scope of the new medicine, see Chapter I of Akif Ercihan Yerlioğlu, 'Paracelsus Goes East: Ottoman "New Medicine" and Its Afterlife'. (PhD. Diss., Harvard University, 2020), 14–69.

54 Yerlioğlu, 16.

55 Akif Ercihan Yerlioğlu, 'Who "Banned" New Medicine in the Ottoman Empire?', *Osmanlı Araştırmaları / The Journal of Ottoman Studies* 63 (2024): 195–225.

56 Yıldırım, 'The Age of the Perplexed: Translating Nature and Bodies Between the Ottoman Empire and Europe, 1650–1730', 177.

57 Duygu Yıldırım, 'Back to the Future: A Belated History of "New" Science in the Ottoman Empire', *BJHS Themes* 9 (2024): 169–182.

require further evidence, as 'new' was a reality for Hayâtîzâde rather than a method of conceptualisation. Similarly, the argument for the shift from Islamic canon medicine would need further proof, as Hayâtîzâde's intention, rather than an explicit and categorical rejection of the old medicine, was to consider a new medical corpus. In turn, the place of the 'old' canon (*kadim*) within new medicine has itself become a recent subject of debate. In their 2023 article, for example, Müstakim Arıcı and Esra Aksoy have advocated a flexible approach towards the concept of new medicine. To them, this flexibility seems to have stemmed from the encompassing and unchanging nature of the old (*kadîm*) in its encounter with the new.[58] In other words, 'new' was a reworking of the old instead of being entirely novel. Despite the nuanced sophistication of these recent discussions, the degree to which change is perceived and acknowledged in this framework has remained ambiguous due to the emphasis placed on the old.

As a result, several characteristics commonly attributed to the new medicine still continue to sustain the impression of an epistemic jump from the old. These include: (1) that the new medicine marked the advent of iatrochemical medicine; (2) that it was Paracelsian;[59] (3) that the arrival of the new medicine eventually marked the waning of old medicine, namely Galenism and humoral pathology.[60] These three points, which emerge from both *The Quintet* and Ibn Sellum's *The Utmost Explanations on the Protection of the Human Body* (*Gayetü'l-Beyan fi Tedbir-i Bedeni'l İnsan*), require further investigation, as the ambiguity of the category of 'New Medicine' at times creates confusion as to whether it represents a reality addressed by the historical actors or is imposed by historians today.

First and foremost, we should keep in mind that the use of complex chemical compounds as therapeutics in the Ottoman lands long preceded the sev-

58 Mustakim Arıcı and Esra Aksoy, 'Tıbb-ı Cedîd Ne Kadar Yenidir? Osmanlı Tıbbında Yenilik Tartışmaları ve Yeninin Mahiyeti (1650–1750)', *Osmanlı Bilimi Araştırmaları / Studies in Ottoman Science* 24, no. 2 (2023): 841–876.

59 It is interesting that such names as Daniel Sennert have been openly associated with Paracelcism in the historiography of Islamic medicine. See Peter Ernst Pormann and Emilie Savage-Smith, *Medieval Islamic Medicine*, The New Edinburgh Islamic Surveys (Edinburgh: Edinburgh University Press, 2007), 171. In reality, Sennert adopted an eclectic approach to Galenism and chemical medicine and openly criticised Paracelsus and his followers for appropriating an older tradition of chemical medicine. A similar tendency is also present in Ottoman historiography. For a specific association of Sennert with Paracelsian drug recipes, see Küçük, *Science without Leisure*, 146.

60 See, for an argument on the waning of humoral pathology in Hayâtîzâde's work, Yıldırım, 'The Age of the Perplexed: Translating Nature and Bodies Between the Ottoman Empire and Europe, 1650–1730', 193.

enteenth century, the most well-known example being Davud al-Antaki in the sixteenth century.[61] Second, Paracelsus was not in fact among the many European doctors that Hayâtîzâde cited. Nor did his predecessor Ibn Sellum cite Paracelsus in his treatise *The Utmost Explanations*.[62] Therefore, a direct Paracelsian influence remains an open question, at least as far as the late seventeenth century is concerned. In addition, it is worth noting that Paracelsus remained quite a contested figure also in sixteenth- and seventeenth-century Europe, as some physicians rejected his sharp criticism of the Ancients.[63] Finally, there was no explicit rejection of humoral theory in favour of a new medicine in Hayâtîzâde's *Quintet*, as he does not seem to move away from either humoral pathological categories or some intrinsically Galenic concepts and practices such as the six non-naturals, bloodletting and therapies. As will be discussed below, a better description of his approach would be a reconciliation of humoral medicine with that of contemporary Latinate medicine, particularly that of the eclectic Daniel Sennert, filtered through the expertise of Hayâtîzâde.

Finally, neither Ibn Sellum nor Hayâtîzâde make any references to *tıbb-ı cedîd* ('new medicine') by name; it seems to be a later conceptualisation. Instead, what they emphasised was change and the need for updated therapies. Therefore, it is more plausible to speak in terms of a transformation noticed by both authors, rather than restricting their mindset to a category we envisage for them. In fact, 'change' or 'new' could well be a concern of an author writing in any period, not only in the late seventeenth century. In this respect, the most crucial task is to recognise such concerns and perceptions of novelty within

61 Antaki is known especially for his recommendation of internal use of mercury before the so-called Paracelsian new medicine. Natalia Bachour, '*Healing with Mercury*: The Uses of Mercury in Arabic Medical Literature', *Asiatische Studien – Études Asiatiques* 69, no. 4 (2015): 854–857; See also Bachour, 'Iatrochemistry and Paracelsism in the Ottoman Empire in the Sixteenth and Seventeenth Centuries', 85.

62 In *The Utmost Explanations*, he only refers to *etibbâ-i kimyevi*: Zekiye Gül Elbir, 'Salih Bin Nasrullah (Ibn Sellum El-Halebi) Gayetü'l-Beyan Fi Tedbiri Bedeni'l-İnsan (Giriş-İnceleme-Metin-Dizin)' (PhD. Diss., Fırat Üniversitesi, 2000), fol. 141^{b}. As for the posthumously completed translations commissioned by Ibn Sellum, Bachour demonstrates that some of them were also not Paracelsian in nature throughout, even when translated from noticeable Paracelsists such as Oswald Crollius: *al-Kimiya al-Malakiyya*, the Arabic translation of the latter's *Basilica chymica*, is not interested in philosophical aspects (i.e., the rejection of humoral pathology) of Paracelcism but only the alchemical methods for remedies. See the conclusion of Bachour, 'Iatrochemistry and Paracelsism in the Ottoman Empire in the Sixteenth and Seventeenth Centuries', 111.

63 For instance, as Dag Nikolaus Hasse has argued, the old medicine was still the preferred practice in sixteenth-century Germany after Paracelsus: Dag Nikolaus Hasse, *Success and Suppression: Arabic Sciences and Philosophy in the Renaissance* (Cambridge, Mass. and London: Harvard University Press, 2016) 19.

Hayâtîzâde's own agenda. To this end, my use of the adjective 'new' simply denotes a novel body of knowledge, namely a number of medical works produced in Europe in the late fifteenth, sixteenth and early seventeenth centuries, of which Hayâtîzâde and his intellectual milieu had a growing awareness. In doing so, my intention is also to consider the possibility that such works may have been already utilised by some Ottoman doctors prior to Ibn Sellum and Hayâtîzâde.

Practical Modes of Textual Production in *The Quintet*

Hayâtîzâde's mode of writing possesses several qualities that are worth noting. To begin with, his texts are based on a systematic analysis of other scholars' works, namely a group of European medical authors from the fifteenth to seventeenth centuries. As we will also see through some examples below, this critical mode of writing provides him with the opportunity to engage with, adopt and reject selectively among a plenitude of ideas. In this way, he also elaborates extensively on pathology, aetiology and symptomatology. Furthermore, he devises his own technique and means of textual investigation, systematically pursuing it in each treatise in *The Quintet*. He conceptualizes certain matters to discuss for each disease. These are: (a) naming, (b) disease aetiology, (c) the reasons for transmission and process of transmission or affliction, (d) symptoms, (e) therapeutic modalities and (f) therapeutic modalities for each symptom. Although the particular aspects can change according to the scope of the subject in each treatise,[64] Hayâtîzâde's handling of each disease remains almost the same, a reflection of his practicality and rationality.

Duygu Yıldırım has argued that Hayâtîzâde utilised a new 'epistemic genre' and observed that, because of his epistolary style, his writing resembled the genre of *Observationes*, the physicians' collections of case histories.[65] But she

64 *Treatise on Hypochondriac Disease* (*Risâle-i illet-i merâkiyye*) is composed of nine chapters; *Treatise on Hypochondriac Melancholy* (*Risâle-i sevda-ı merâkiyye*) is composed of seven chapters; *Treatise on Syphilis* (*Risâle-i maraz-ı efrenci*) is composed of twenty-five chapters; *Treatise on Plica Disease* (*Risâle-i pelika*) is composed of eight chapters; and *Treatise on the Pestilential Fever* (*Risâle-i hummâ-yı redîye*) is composed of thirteen chapters.

65 Yıldırım, 'The Age of the Perplexed: Translating Nature and Bodies Between the Ottoman Empire and Europe, 1650–1730', 177. Yıldırım's point of reference here is Gianna Pomata's two articles: see Gianna Pomata, 'Sharing Cases: The Observationes in Early Modern Medicine', *Early Science and Medicine* 15, no. 3 (2010): 193–236; Gianna Pomata, 'The Medical Case Narrative: Distant Reading of an Epistemic Genre', *Literature and Medicine* 32, no. 1 (2014): 1–23.

does not elaborate further on this resemblance in her dissertation. However, Hayâtîzâde's references to Amatus Lusitanus (*Amato*) (1511–1568), Ercole Sassonia (*Saşuniye*) (1551–1607), Petrus Forestus (*Forusto*) (1521–1597) and Epifanio Ferdinando (*Ferdinandus*) (1569–1638) demonstrate that the connection goes beyond mere stylistic resemblances, as these four authors participated with their case histories in the creation of the genre of *Observationes*.[66] As I discuss below, Hayâtîzâde referred to such clinical observations particularly in his account of syphilis because he admittedly lacked clinical experience in some aspects of its treatment. Thus, whether Hayâtîzâde aimed at reinventing a new medical genre in the Ottoman context remains a topic to explore further, but there is no doubt that his practical modes of thinking and writing enabled him to approach his subject from multiple points of view.

Each treatise in *The Quintet* starts with a chapter that discusses the reason for the name (*vech-i tesmiye*) of the disease in question and aims to inform the reader about the name and the origins of that disease. The interest in disease nomenclature was not entirely a new phenomenon in Ottoman medical writing. One of Hayâtîzâde's predecessors Sakızlı İsa Efendi (d. 1647) had also paid careful attention to disease names in his *Cure for Diseases* (*Devâü'l-Emrâz*) and listed the names of every ailment in different languages including Greek, Latin, Persian, Arabic and so on.[67] Nevertheless, Hayâtîzâde recounted also stories of how a particular disease came to be named. For example, the first section of the *Treatise on Syphilis* (*Risâle-i Marâz-ı Efrencî*) narrates, referring to Daniel Sennert, that some Spanish vessels sailed westward in 1493 and arrived in the Indies, that is the New World (*Yeni Dünyâ*). Following their conquests, they noticed that the inhabitants were afflicted with this disease, which they eventually brought back to their homeland. After the Spanish, the French and the Dutch travelled there and returned with the same malady. As for the naming, all three people – Spanish, Dutch, and French – blamed the others, but in the end, it came to be known as the French disease. In the Ottoman lands, however, it is called the *freng marazı*, the 'European disease'.[68]

One can assume that Hayâtîzâde sought to introduce his audience to these 'new' diseases, namely syphilis and plica, by starting with their names because people probably did not know many aspects of these diseases, whose names

66 For Forestus and Lusitanus see 218 (footnote 59); for Sassonia see 222 (footnote 69), for Ferdinando see 229–230; Pomata, 'Sharing Cases'.

67 See the beginning of each chapter (*bâb*) in İsa Efendi *Devâü'l-Emrâz,* Kütahya Zeytinoğlu Halk Kütüphanesi, 401.

68 Hayatizade Mustafa Feyzi Efendi, 'Resâilü'l-Müşfiyye Li'l-Emrâzi'l-Müşkile', 1693, fols. 49[b]–50[a]. Hayâtîzâde provides the date in the Gregorian calendar.

derived from an unfamiliar language. For instance, the first chapter (*fasl-ı evvel*) of the *Treatise on Plica Polonica* (*Risâle-i Pelika Polonika*) elaborates on both of the two words in the title. Hayâtîzâde writes that plica means 'matted' hair whereas Polonica means the land of the Polish people in Latin. 'In short, it means a hair disease in the Polish lands.'[69]

Since Hayâtîzâde usually continues with a new chapter on the essence (*keyfiyet ve mahiyet*) of each disease, it seems highly likely that there was an interest in the nature and origins of these diseases among various audiences. In this part, the author not only discusses the anatomical and environmental sources of these diseases but also colourfully analyses the debates between his sources. In this regard, the essence section of the *Treatise on Syphilis* (*Risâle-i Marâz-ı Efrencî*) presents a highly intriguing example in which Hayâtîzâde's three sources (Daniel Sennert, Luis Mercado, and Abraham Zacutus Lusitanus) agree that this disease is caused by an unknown poison (*semmiyet-i hafiyye*) and what matters is its quality (*keyfîyyet*), not its quantity (*kemmiyyet*). However, as Hayâtîzâde shows later in the section on remedies of the same treatise, these authors disagree on the methods to adopt to cure syphilis. Hayâtîzâde manages to uniquely position himself as an 'objective' critic of the literature with which he engages.

This chapter is typically followed by a new one on disease aetiology (*esbâb-ı maraz*). Hayâtîzâde's perception of disease aetiology and his theorisation fuel our knowledge of how disease affliction and transmission (*sirâyet*) were understood among learned physicians in the late seventeenth-century Ottoman capital. This part on disease aetiology is usually followed by or combined with another chapter on how it is spread and caught (*sirâyet*). For instance, Hayâtîzâde describes the transmissibility of syphilis based on the information he derived from Luis Mercado where he basically notes it may be transmitted via contact with the afflicted person through being exposed to their breath, using the same utensils, dressing in their clothes and/or sleeping on the same pillow and bedsheet.[70] Comparative means were also considered in explaining disease affliction. For instance, pestilential fever (*humma-yı vebaîyye*), whose causes were considered 'hidden', was from the category of transmissible dis-

69 'Mâlum ola ki bu maraza pelika polonika tesmiye eylediler. Vech-i budur ki Latin lisânında lafz-ı pelika buruşmış saç demektir. Polonika Leh vilayeti dimekdir. Hâsıl-ı mânâ Leh diyarında vâki saç marazı dimekdür' ('Be it known that they named this disease plica polonica. It is because plica means matted hair in Latin. [And] Polonica means the Polish land. In short, it is a hair disease that is from the Polish land.'), *Hayâtîzâde Mustafa Feyzi Efendi, 'Resâilü'l-Müşfiyye Li'l-Emrâzi'l-Müşkile, 1693, fols. 96*[b]*–97*[a].

70 Hayâtîzâde Mustafa Feyzi Efendi, fol. 54[b].

eases (*emrâz-ı vafîde*) similar to the bubonic plague (*taûn*).[71] Yet it did not afflict in a similar manner to the bubonic plague. The etiological comparison between pestilential fever (*hummâ -yı vebâiyye*) and the bubonic plague demonstrates that other possible means of transmission were being scrutinised and even conceptualised by Hayâtîzâde and his contemporaries.

Following this part, Hayâtîzâde usually lists the symptoms (*âraz*) in a separate chapter. A further chapter, or chapters, discusses the remedial and prophylactic measures, providing detailed recipes for medicines. The last section of each treatise typically elaborates on the methods of treatment for each specific symptom that follows the onset of the disease. Even though each treatise is basically comprised of these chapters, there may be some additional ones specific to each treatise depending on the unusual characteristics of the disease and/or the need for further warning about their possible modalities of treatment. For instance, the author adds specific chapters on the rules to consider while using mercury oil (*duhn-ı zeybak*) as it can be extremely dangerous for both the patient and the practitioner.[72] In addition, Hayâtîzâde wrote special chapters called *takdimetü'l-mârife*. Elif Nur Önder has established that Hayâtîzâde refers to the *Book of Prognostics* from the Hippocratic corpus with the phrase *takdimetü'l-mârife*.[73]

Yıldırım discusses how Hayâtîzâde's pragmatic textual preferences stand in stark contrast to the previous characteristics of the Ottoman medical corpus.[74] Indeed, Hayâtîzâde's interest in the individual experience of a patient and case histories and his practical modes of textuality demonstrate that he wanted to supersede existing theoretical limits. However, the argument that this was a novel system of writing that broke from the previous corpus requires further evidence. One needs to remain cautious in assessing particularly whether this extended focus on pathology was a novelty or an epistemic shift. Looking at previous examples of medical books, one may indeed notice a pattern of shorter pathological explanations. For instance, the limited scope of aetiological explanations in Ibn Sellum's *Utmost Explanations* might lead us to think that Hayâtîzâde's extensive pathological commentaries were a novelty.

71 Hayâtîzâde Mustafa Feyzi Efendi, fol. 114[b].

72 See chapter 20 of the Treatise on Syphilis: 'Yirminci Fasıl: Duhn-ı zeybâk tılâ olmazdan mukaddem lâzım olan şurûtı beyân ider' ('[This] conveys the rules to follow prior to application of mercury oil'), see Hayatizade Mustafa Feyzi Efendi, fol. 82[a].

73 See footnote 20 in Önder, 'Hayatizade'nin İllet-i Merakiyye ve Sevda-Yı Merakiyye Risalelerinin Tıptaki Ahlat-ı Erbaa Düşüncesine Göre İncelenmesi ve Transkripsiyonlu MetnI', 7.

74 Duygu Yıldırım Yıldırım, 'The Age of the Perplexed: Translating Nature and Bodies Between the Ottoman Empire and Europe, 1650–1730', 195–196.

Yet this assumption is undermined if we look at one of Ibn Sellum's predecessors, Sakızlı İsa Efendi (d. 1647), whose *The Cure for Diseases* (*Devâü'l-Emrâz*) showed almost as much interest in pathology as Hayâtîzâde did.[75]

'Latinate Physicians' between Theory and Practice

The Quintet was formed in conversation with the works of around twenty-five Renaissance physicians, 'Latinate' physicians (*etibbâ-yı Lâtin*) in Hayâtîzâde's words, who were intellectually active at some point between the late fifteenth and early seventeenth centuries. Arguably, the majority of these physicians were present at Padua Medical School, the most prestigious school of medicine in Europe in medieval and early modern times.[76] Some of these physicians Hayâtîzâde cites are: Antonio Benivieni (*Anton Benüvinü*) (1443–1502), Johannes Baptista Mantonus (*Mentona*) (1498–1551), Girolamo Fracastoro (*Frokastor*) (1478–1553), Jean Fernel (*Fernelyos*) (1506–1558), Gerolamo Cardano (*Kardano*) (1501–1576), Nicolás Monardes (*Nikolaus Monardus*) (1512–1588), Amatus Lusitanus (*Amato*) (1511–1568), Ercole Sassonia (*Saşuniye*) (1551–1607), Luis Merkado (*Merkado*) (1525–1611), Felix Platter (*Plattor*) (1536–1614), Wilhelm Fabry (*Faberiso*)[77] (1560–1634), Petrus Forestus (*Forusto*) (1521–1597), Daniel Sennert (*Senertos*) (1572–1637), Epifanio Ferdinando (*Ferdinandus*) (1569–1638), Abraham Zacutus Lusitanus (*Zakota*) (1575–1642).[78]

This was a diverse group of medical scholars.[79] Some of them were famous botanists: Nicolás Monardes who introduced plants from the New World to Europe in the late sixteenth century is the most significant name that falls

75 Two chapters of my dissertation focus on İsa Efendi's *The Cure for Diseases*, examining the efforts to medicalise certain conditions such as appetite. For a general overview of *The Cure*, see Ahmet Akdağ, 'Sakızlı İsa Efendi'nin Deva-i Emraz İsimli Eseri', in *Sakızlı İsa Efendi ve Nizamü'l Edviye'si Üzerine İncelemeler*, ed. Sibel Murad and Mücahit Kaçar (İstanbul: Dün Bugün Yarın Yayınları, 2021).

76 For the significance of the Padua Medical School in medieval and early modern Europe, see Fabio Zampieri, 'The University of Padua Medical School from the Origins to the Early Modern Time: A Historical Overview', in *Scientiae in the History of Medicine*, ed. Fabrizio Baldassarri and Fabio Zampieri (Roma: 'L'Erma' di Bretschneider, 2021), 23–68.

77 This physician called *Faberiso* could also be Hieronymus Fabricius, the famous anatomist from Padua.

78 Some of the names are yet to be identified such as *Diyoriyo* and *Canfort*. Hayâtîzâde's versions are in brackets.

79 A more detailed discussion of the physicians cited in *The Quintet* will come in my dissertation.

within that category.[80] Recent research has suggested that Monardes's work was read in Northwest Africa at least from the late sixteenth century onwards through Arabic translations.[81] There is also strong evidence that Turkish translations were available as early as the early seventeenth century.[82] Yet Hayâtîzâde's source of reference must have been a Latin translation of Monardes.[83] As discussed above, some of these names are known to have spearheaded the advent of a patient-oriented approach in early modern European medicine.

Hayâtîzâde's innovative and revisionist tendencies are evident in the book, as he writes at the beginning of *The Quintet* that the Arab and Persian physicians did not understand the nature of the five diseases in question and thus were incapable of finding the right treatment for them. He argues that hypochondria and melancholy hypochondria were regarded as the same disease in this corpus since they followed the medicine of the Ancients, that is the medicine of Hippocrates and Galen. As the latter did not differentiate between the two diseases, therefore, neither did the Arabs and Persians. In addition, the pestilential fever studied in the fifth chapter was known only by its name in the preceding medical literature. Thus, the Arab and Persian physicians were likewise unable to diagnose and treat this condition. According to Hayâtîzâde, the last two, that is plica and syphilis, had never been mentioned in the

80 For the reception of Monardes in European vernacular languages, see Alisha Rankin, 'New World Drugs and the Archive of Practice: Translating Nicolás Monardes in Early Modern Europe', *Osiris* 37 (2022): 67–88.

81 See Aslıhan Gürbüzel's discussion on the translations conducted by Ibn Jani in Aslıhan Gürbüzel, 'From New Spain to Damascus: Ottoman Religious Authorities and the Making of Medical Knowledge on Tobacco', *Early Science and Medicine* 26, no. 5–6 (2021): 569–574.

82 Alexandar Shopov has discussed *The Treatise on Tobacco* (*tütün risâlesi*) that was translated into Turkish by a certain Mustafa Efendi in 1608 from an Arabic treatise by a Moroccan scholar named Mansur. Shopov has pointed out the similarities and differences of both the Turkish translation and the Arabic original with Monardes's discussion of tobacco. See Aleksandar Shopov, 'Between the Pen and the Fields: Books on Farming, Changing Land Regimes, and Urban Agriculture in the Ottoman Eastern Mediterranean ca. 1500–1700' (PhD Diss., Harvard University, 2016), 270–271.

83 Hayâtîzâde directly references Monardes in the *Treatise on Syphilis*, saying 'Mâlum ola ki Nikolaus Monardus nâm tabib yeni dünyada vâki olan eczânın müfredâtını beyan eylemişdür.' ('Be it known that the physician named Nikolaus Monardus elaborated on the pharmacopeia of the New World'), Hayatizade Mustafa Feyzi Efendi, 'Resâilü'l-Müşfiyye Li'l-Emrâzi'l-MüşKiley, 1693, fol. 72[a]. Since we know that Hayâtîzâde was using books in Latin, he must have been using the Latin translations of Monardes produced by the Dutch physician Carolus Clusius. For more on Monardes's translations in Latin, see Rankin, 'New World Drugs and the Archive of Practice', 68.

works of the Arabs and Persians.[84] In other words, they were new diseases (*marâzân-ı cedîdân*). Yet, Hayâtîzâde was not the first to treat them: Arab physicians such as Al-Antabî and 'Imâd al-Dîn Mahmûd b. Mas'ûd b. Mahmûd al-Tabîb had noted the internal use of mercury in treating syphilis already in late sixteenth-century Egypt and the Arab world.[85] Both were also mentioned by Hayâtîzâde's predecessor Ibn Sellum (d. 1669) in his *The Utmost Explanations*.[86] Syphilis was also on the agenda of Sakızlı İsa Efendi (d. 1649), the predecessor of both Ibn Sellum and Hayâtîzâde and Chief Physician of Sultan Ibrahim.[87]

Convinced of the superiority of the writings of the Latinate physicians on the diagnosis and treatment of these diseases, Hayâtizâde emphasises the urgent need to update existing knowledge in the field. He states that he undertook the task of compiling this work by relying on 'the books and treatises of the ancients as well as the publications of some prominent Latinate physicians' that he had in his possession at that time.[88] The ancients (*kudemât*) referred to here are none other than Hippocrates (*İbokrat*) and Galen (*Câlînûs*), whose medical corpus had informed Islamic medicine over the centuries. The existence of two different medical traditions in *The Quintet* arguably indicates that Hayâtîzâde did not totally reject the ancients to adopt a new medical framework. Rather, his perception aligned well with the medical currents of Europe in the seventeenth century, as this era was indisputably marked by the development of diverse medical perspectives in Europe and witnessed extensive efforts to reconcile the Galenic framework with that of chemical medicine.

Daniel Sennert (1572–1637), the German physician and professor of medicine, is a good example of this attitude, and his name is especially relevant here since he is arguably the most cited name in *The Quintet*. However, contrary to his representation in the historiographies of Islamic and Ottoman medicine (and mostly with reference to Ibn Sellum), Daniel Sennert never identified himself as a follower of Paracelsus. In fact, he elaborated on this question in his

84 Hayatizade Mustafa Feyzi Efendi, 'Resâilü'l-Müşfiyye Li'l-Emrâzi'l-Müşkile', 1693, fols. 1[b]–2[a].

85 Bachour, *'Healing with Mercury'*, 851–854. Also see Bachour, 'Iatrochemistry and Paracelsism in the Ottoman Empire in the Sixteenth and Seventeenth Centuries', 90–91.

86 For a transcription of *Gayetü'l-Beyân*, Elbir, 'Salih Bin Nasrullah (Ibn Sellum El-Halebi) Gayetü'l-Beyan Fi Tedbiri Bedeni'l-İnsan (Giriş-İnceleme-Metin-Dizi)', 287[a].

87 Sakızlı İsa Efendi, 'Devaü'l-Emrâz', fols. 321[b], 322[a], 322[b], Kütahya Zeytinoğlu Halk Kütüphanesi.

88 Hayâtîzâde Mustafa Feyzi Efendi, 'Resâilü'l-Müşfiyye Li'l-Emrâzi'l-Müşkile', 1693, Süleymaniye Kütüphanesi, Selim Ağa Koleksiyonu 868, fol. 1[b].

De Chymicorum (1619),[89] in which he devoted a section to Paracelsus and told the interesting story of the latter's life, including his iconoclastic criticism of the medicine of Galen and Avicenna and the medical establishment of his day such as the medical practitioners in Paris, Padua and Montpellier.[90] According to Sennert, it was almost disrespectful for Paracelsus and his followers to boastfully attribute the title of 'chymists' to themselves, because 'chymistry' was used before Paracelsus.[91] Emily Michael has argued that Sennert was a transitional medical figure in seventeenth-century Europe who brought Aristotelian norms together with the new chemical medicine. In this way, according to Michael, his medicine can only be characterised by the term 'Latin pluralism'.[92] This pluralism understandably created an environment in which scholarly discussions increased and resulted in a considerable number of controversies. Hayâtîzâde, an attentive reader of the Latinate physicians, colourfully narrates these controversies. Textual translation *per se* does not define Hayâtîzâde's engagement with this broad group scholars. He, in a way, reconciles disparate branches of medical knowledge at a theoretical level; even if he thinks that the theories of ancient physicians (*kudemât*) are no longer enough to explain certain disease conditions, he unhesitatingly remains Galenic, not only in his medical framework (i.e., humoral pathology) but also in his therapeutics. But he also acknowledges change. His understanding of change, defined and discussed under the rubric 'new chemical medicine' by several scholars, seems to be mostly driven by Columbian Exchange rather than a sharp interest in chemical medicine. In that sense, his tendencies reflect the general trends of European medicine at that time. Hayâtîzâde's scholarly scrutiny, fuelled by his eagerness to test 'new' knowledge and combine it with local knowledge whenever possible, resulted in an Ottomanisation of Latinate medicine.

Scholarly Scrutiny, Experience and Experimentalism

Hayâtîzâde's scholarly scrutiny in *The Quintet* enabled him to render knowledge meaningful in the local context in both theoretical and practical per-

89 I use the shortened English translation, *Chymistry Made Easie and Useful Or The Agreement and Disagreement of the Chymists and Galenists*, trans. Nicholas Culpeper and Abdiah Cole (London: Peter Cole, 1662).

90 Daniel Sennertus, *Chymistry Made Easie and Useful*, 14.

91 Sennertus, *Chymistry Made Easie and Useful*, 15.

92 Emily Michael, 'Daniel Sennert on Matter and Form: At the Juncture of the Old and the New', *Early Science and Medicine* 2, no. 3 (1997): 273.

spectives. This process of Ottomanisation of Renaissance medical knowledge was only possible through Hayâtîzâde's observations, experiences and experimentalism. Here I will present some examples of such intellectual intervention by Hayâtîzâde. But first, I would like to briefly discuss what I mean by scholarly scrutiny. At a theoretical level, scholarly scrutiny denotes Hayâtîzâde's eagerness to critically engage with the ideas of Renaissance physicians. For instance, he brilliantly discusses the differing opinions of Gerolamo Cardano (*Kardano*) (1501–1576), Felix Platter (*Plattor*) (1536–1614) and Sennert on how mercury (*zeybak*) helps to cure syphilis. Such a critical engagement was significant particularly because he admitted that he lacked clinical experience in using mercury.[93] Therefore, this discussion would provide *a priori* assistance to both him and his readership. Yet, in other instances, scholarly scrutiny was not limited to theoretical discussion; observation was present as an empirical modality in many cases to allow Hayâtîzâde to bring his expertise into play. For instance, while discussing the symptoms of melancholy, Hayâtîzâde cautions his reader not to mistake this specific disease for the syndrome of renal calculi. He supports this argument by referring to Sennert, who had noted that this particular symptom of melancholy differs from the renal calculi, in which it is possible to detect calculary remnants in the patient's urinary culture.[94] At this point, Hayâtîzâde intervenes, adding that he himself had observed a difference in the calculary samples of the two diseases: 'lâkin bu fakir bögreklerden gelen kum ile merâkiyyeden gelen kumun mâbeynlerin fark eyledik' ('but I, this humble one, observed that the characteristics of the gravels coming from hypochondria and the kidneys are different').[95] He elaborates on his findings at the end of the chapter.[96]

However, Hayâtîzâde's critical engagement does not end here, as he further brings to play another empirical modality, namely experience: For example, he reveals at some point that he also suffered from hypochondria and regularly experienced at least one symptom of the disease, namely heart palpitations. He adds that he used a specific kind of oil extracted from citron zest and cinnamon and benefitted from it:

---

93 'Amma bu fakir zikr olunan etibbânın tedbirlerinde ekseriya hata olmayub isabetlerine şehadet eyler iken duhn-ı mezbûrı maraz-ı efrence müddet-i ömrümde istimâl eylemek vâki olmadı.' ('I confirm that all of the methods suggested by these physicians are not erroneous, while I have never had a chance in my life to apply the mentioned oil [of mercury] to the European disease'), Hayatizade Mustafa Feyzi Efendi, 'Resâilü'l-Müşfiyye Li'l-Emrâzi'l-MüşKiley, 1693, fol. 77ª.

94 Hayâtîzâde Mustafa Feyzi Efendi, fol. 12ª.

95 Hayâtîzâde Mustafa Feyzi Efendi, fol. 12ᵇ.

96 Hayâtîzâde Mustafa Feyzi Efendi, fols. 14ᵇ–15ª.

> Bu fakir maraz-ı merakiyyeye mübtela olmağla hafakan arız oldukda tecrübe etdim, azim nefin müşahede eyledim. Senertos hekim sebebi rih olan hafakanının muâlecesinde hükema-i Efrenc'in taktir ile ağaç kavunı kabığından ve darçından istihraç eyledikleri yağları gâyetü'l-gâye mâkul gördüm dedi. Fakir dahi bu yağları Frengistan'dan İstanbul'da Galata'ya gelmek ile alub kendü nefsimde tecrübe eyledim. Bi-lütfullah-ı teâla nefini müşahede eyledim.[97]
>
> (When I, this humble one, was afflicted with the disease of hypochondria, I experimented with this and observed excellent help. Sennert the Physician said he witnessed that the oils extracted from the zest of citron as well as cinnamon were used by Latin doctors and are beneficial for the treatment of the palpitations caused by rheumatism. Even I, this humble one, bought these oils when they were brought to Galata in Istanbul from Europe and experimented with them on myself. By the providence of God, I observed that they were beneficial.)

This remark, in which he refers to himself as both physician and patient, makes quite an intriguing case. As Andrew Scull points out in his seminal book *Madness in Civilization: From the Bible to Freud, from the Madhouse to Modern Medicine*, melancholy was regarded as an affliction of 'the scholar and the man of genius'.[98] In a similar manner, Robert Burton, the author of *The Anatomy of Melancholy*, suffered from melancholy and started writing *The Anatomy* as a homeopathic means that he believed would cure his affliction.[99] Hayâtîzâde's affliction with hypochondria allows us to situate him among such examples in the early modern period.

Finally, the whole process of reconciliation and vernacularisation was completed using a form of experimentalism. His long discussions of remedies for syphilis, for instance, included medicinal substances from the New World such as guaiac wood, saparina and sassafras.[100] His source of reference was

---

97 Hayâtîzâde Mustafa Feyzi Efendi, 'Resâilü'l-Müşfiyye Li'l-Emrâzi'l-Müşkile', 1693, fols. 29b–30a.

98 Andrew T. Scull, *Madness in Civilization: A Cultural History of Insanity, from the Bible to Freud, from the Madhouse to Modern Medicine*, 2nd ed (Princeton: Princeton University Press, 2015), 93.

99 Angus Gowland, *The Worlds of Renaissance Melancholy: Robert Burton in Context* (Cambridge: Cambridge University Press, 2006), 2–3.

100 For guaiac wood, see fols. 63b–68b; for saparina, see 68b–70a; for sassafras, see 72a–72b in Hayatizade Mustafa Feyzi Efendi, 'Resâilü'l-Müşfiyye Li'l-Emrâzi'l-Müşkile', 1693.

Monardes.[101] He mentions that the guaiac wood, that originates in Peru, came to be known as *peygamber ağacı* (the prophet's wood) in Turkish.[102] Such naming practices of novel substances indicate a collective eagerness to vernacularise new substances in the Ottoman context.[103] Furthermore, the possible cures for syphilis reconcile the local plants and the new ones. Thus, caudles prepared with local species such as *Amasya eriği*, a specific type of plum that grows in the Anatolian province of Amasya, constitute the primary steps in the treatment process.[104] In a similar manner, Galenic therapies also remain available along with new ones. For example, in case there is strong fever in the body, the guaiac wood should be strictly avoided. The cure recommend for this fever was the Galenic soothing ointment (*Calinus'un soğudıcı merhemin*).[105]

All in all, the medical knowledge produced in the late-seventeenth-century Ottoman court was the result of a systematic interaction with the works of Renaissance physicians in Europe. This was enabled by the recruitment of physicians with knowledge of the Latin language in the Ottoman court. The expertise of the court physicians, which included competence in experimental observation, meant that the transfer process was not one of mere translation. Instead, as the example of Hayâtîzâde demonstrates, they adopted and/or rejected aspects of this corpus based on their expertise, at times contributing to the construction of a new medical knowledge, that was shaped by the global currents of the day.

Acknowledgements

This chapter, which presents some of the preliminary results of my ongoing doctoral research, is a revised version of a research paper submitted to the History Department at the European University Institute in June 2021. I would

101 Hayatizade Mustafa Feyzi Efendi, 72[a].

102 Hayatizade Mustafa Feyzi Efendi, fol. 63[b].

103 The name, the prophet's wood, must have been created and popularised in the late seventeenth century. In the 1640s, İsa Efendi refers to guaiac as *'gayaku'*, and a reader of Kütahya manuscript notes in the marginalia later: *'guaiac is the prophet's wood'* (*gayaku peygamber ağacıdır*), 'Devaü'l-Emrâz', fol. 322[a]. This also raises the question as to the temporal scope of the 'new', as it shows that the existence of such plants and their knowledge in the Ottoman world preceded Hayâtîzâde and Ibn Sellum. My forthcoming dissertation discusses this issue in detail.

104 For a caudle recipe that includes Amasya plum and other local plants, see Hayatizade Mustafa Feyzi Efendi, 'Resâilü'l-Müşfiyye Li'l-Emrâzi'l-Müşkile', 1693, fol. 62[b].

105 Hayatizade Mustafa Feyzi Efendi, fol. 68[b].

like to thank Prof Ann Thomson and Prof Giancarlo Casale for inviting me to share my research in the PIMo Workshop 'Translators, Translations and Mobility in the Mediterranean' held in Florence in February 2022. My thanks are also due to Jan Becker and Silvia Maria Marchiori for the generous discussions on the topic of vernacular medicine in preparation for our mutual panel 'Writing Early Modern Vernacular Medicine' in the British Society for the History of Science Postgraduate Conference held on 1 April 2023. I am grateful to Prof Nükhet Varlık for her extensive comments and feedback. Finally, I thank Dr İsa Uğurlu for sharing with me his findings from the probate inventories of Hayâtîzâde's descendants.

CHAPTER 7

'Just Learning the Languages of India …'

Worlds of Translation in Daniello Bartoli's Missione al Gran Mogor *(1663)*

Angelo Cattaneo

Introduction

This chapter explores the cultural and linguistic dimensions of Daniello Bartoli's *Missione al Gran Mogor* (1663), a hagiographic and historiographical account of the Jesuit mission to the Mughal court of Emperor Akbar the Great (1542–1605), led by Rodolfo Acquaviva S.J. (1550–1583) in the years right before his martyrdom. By drawing on the extensive archival material preserved in the *Archivum historicum* of the Society of Jesus in Rome, Bartoli provides insight into the Jesuits' early efforts to engage with the religious, linguistic and political complexity of the Mughal Empire and offers a rich account of how linguistic practices – such as the study of Persian and the translation of Christian texts – had become central to the missionary enterprise.

The chapter is structured around two main interwoven sections. After outlining Acquaviva's stay at Fatehpur Sikri, it focuses on a crucial dimension of the mission: the linguistic and translational practices employed by the missionaries. Through selected passages from Bartoli's narrative, it delves into the cultural and religious translations carried out at Akbar's court, particularly through the collaboration between Acquaviva and the Persian courtier Abu 'l-Fadl (1511–1602) in translating parts of the Gospels into Persian. These interactions, while rooted in different ideological aims, represent powerful moments of dialogue and mutual influence. From the Jesuit perspective, learning local languages – especially Persian – was seen as essential to effective negotiation of the both the Jesuit presence and proselytism and illustrates how linguistic and translational practices had become central to missionary efforts.

The second section situates these practices within three coeval interconnected paradigms: the Portuguese imperial model, which linked language dissemination and imposition to territorial expansion and religious conversion, as theorised in João de Barros's grammars and chronicles around 1540, building on Antonio de Nebrija's motto 'the language, companion of the empire'; the Catholic missionary-linguistic tradition, which involved the systematic codific-

 | DOI:10.1163/9789004731806_009

ation and translation of previously unknown languages across the Americas, Africa and Asia as essential practices of missionary outreach; and the Mughal model, which envisioned Persian as a *lingua franca* for a culturally and linguistically plural empire where translation served to harmonise diverse religious and intellectual traditions under Mughal imperial patronage.

By foregrounding these paradigms, the essay argues that language learning and translation were privileged lenses through which early modern actors – Jesuit missionaries, Portuguese and Mughal elites alike – conceived, contested and enacted broader projects of cultural authority, religious identity and political power. Bartoli's text, read in this light, reveals the intricate and often ambivalent processes of mutual accommodation, resistance and transformation in the global missions of early modern Catholicism, even amid deep ideological and geopolitical asymmetries.

Bartoli's *Mission to the Great Mogor*: Rodolfo Acquaviva S.J. at the Mughal Court

In 1663, some eighty years after the violent death of Rodolfo Acquaviva S.J. and his four confreres[1] near the village of Cuncolim, about forty kilometers south of Goa, Daniello Bartoli S.J., the official historian of the Society of Jesus, published a work entitled *Missione al Gran Mogor del P. Ridolfo Acquaviva della Compagnia di Giesù, Sua Vita e Morte E d'altri quattro Compagni uccisi in odio della Fede in Salsete di Goa* (Mission to the Gran Mogor by Fr. Ridolfo Acquaviva of the Society of Jesus, his life and death and that of four other companions killed in hatred of the Faith in Salcete near Goa). First published in Rome by the Varese Press in 1663, the work was added as an appendix to the third edition of the volume on Asia of Bartoli's *Istoria della Compagnia di Gesù* (History of the Society of Jesus). Reprinted in 1714 and then edited several times in the nineteenth and twentieth centuries, the *Missione* was intended to extol the peculiar qualities and the austere and devout existence of Rodolfo, as well as his martyrdom and the 'esteem of holiness' he had acquired both in missionary contexts and in Europe, with the ultimate goal of promoting his beatification, together with the four brothers who perished with him.[2] In doing so, Bartoli's

1 Alfonso Pacheco (1553–1583), of Spanish origin, Pietro Berno da Ascona (1553–1583) and the Portuguese António Francisco (1583) and Francisco Aranha (1551–1583).

2 See in particular the following chapters of the *Missione al Gran Mogor del P. Ridolfo Acquaviva della Compagnia di Giesù*: 'Universale stima di santità in che era il P. Ridolfo' (143–147); 'Singolari virtù del P. Ridolfo ancor giovanetto e secolare. Dono d'oratione, con rapimento

text also contextualised the martyrdom of Rodolfo Acquaviva and his companions against the background of the earlier mission he led to the Mughal court in Fatehpur Sikri.[3]

Based on the documentation preserved in the *Archivum historicum* of the Society of Jesus in Rome, which Bartoli began studying as early as 1648 as he prepared his massive *Istoria della Compagnia di Gesù* (Rome, 1653–1673),[4] the volume tells the story of the mission to the court of the Mughal *Padshah* (the *Great Mogor* of the title) Muhammad Abū'l-Fath Gialāl al-dīn, better known as Akbar (The Great, 1542–1605), undertaken beginning in February, 1580. Accompanying Acquaviva on this earlier mission were two other confreres: Antoni de Montserrat (1536–1600), of Catalan descent, who authored his own volume about the mission, and a more obscure figure, Francisco Henriques (1538–1597), a Persian-speaking Muslim who, following his encounter with the Portuguese in Hormuz, had converted to Christianity. Upon arriving in Goa, Henriques had been admitted to the Society of Jesus and, according to Bartoli, it was his knowledge of Persian as well as his devotion to the Christian faith that had prompted his superiors to choose him for the mission to Akbar's court.[5]

dell'anima in Dio. L'honestà prouìvata a un pericoloso cimento. Sviscerata carità verso i poveri' (147–160); 'Chiamato da Dio alla Compagnia di Giesù, vince i lunghi, e forti contrasti de' suoi parenti, e v'entra', 160–172; 'Qual fosse Ridolfo già Religioso. Rari esempi della sua humiltà, e somma Povertà. Austero seco medesimo. Amabilissimo ad ogni altro' (172–185); 'Perfettione della sua Ubbidienza. Si apporta una sua lettera, piena di nobilissimi sentimenti, che mostrano la sublimità del suo spirito' (185–193); 'Simiglianza di spirito fra i Padri Ridolfo Aquaviva, e Alfonso Paceco. Sentimento del Visitatore Alessandro Valegnani sopra la lor morte. Rare parte di natura, e di virtù nel Paceco' (193–202); 'La Missione dell'India impetrata da Dio con ispecial maniera al P. Paceco. Suoi meriti colla Fede in quelle parti. Particolarità notabile, nella sua morte. Honori fattigli da' Signori della sua Casa' (202–209); 'Contezze della vita e virtù del P. Pietro Berno' (209–215); 'Del P. Antonio Franceschi, e del F. Francesco Aragna' (215–218). The long process of beatification began in 1598 and ended on 30 April 1893 during the pontificate of Leo XIII. For the history of the beatification of Acquaviva and his confreres, see G. De Filippo, *Beato Rodolfo Acquaviva S.I.* (Teramo: Edigrafital, 1999), 101–103; Youri Martini, *Akbar e i gesuiti. Missionari cristiani alla corte del Gran Moghul* (Trapani: Il Pozzo di Giacobbe, 2018), 56.

3 See Bruno Basile's introduction to Daniello Bartoli, *Missione al Gran Mogòr*; ed. Bruno Basile (Rome: Salerno, 1998), 7–20.

4 The *Istoria* was compiled and published by Bartoli over the course of two decades: *Asia* from 1653 (in eight books, with the addition as an appendix of *La missione al Gran Mogor* in 1667), *Japan* in 1660 (in five books), *China* in 1663 (in four books), *England* in 1667 (in six books) and finally *Italy* in 1673 (in four books). See Daniello Bartoli, *Dell'historia della Compagnia di Giesu: l'Asia*, ed. Umberto Grassi and Elisa Frei (Turin: Giulio Einaudi, 2019).

5 'Ma in fine, dopo alquanti dì, cadde la sorte sopra i' PP. Ridolfo Aquaviva, e Antonio Monserrato, a' quali piacque aggiungere un terzo, già per nascimento, d'origine, e di setta moresco, poi

The mission, urged by the *Padshah* himself, was part of Akbar's policy of negotiating, within the kingdom, with Islamic and Hindu religious authorities and outside, with the Portuguese in Cochim and Goa. The missionaries were invited to join Akbar at Fatehpur Sikri, in the vicinity of Agra, a city rebuilt by Akbar and declared the Mughal capital in 1569 at the birth of Akbar's son Salim, who would later succeed his father as *Padshah* in 1605 under the royal name of Jahangir.

As is widely known, the Jesuit mission was organised following the visit of a Mughal ambassador, accompanied by an Armenian interpreter, to Portuguese Goa in September 1579. On this occasion, Akbar had formally requested that Dom Luís de Ataíde, (viceroy of the *Estado da Índia* from 1568 to 1671) send to the Mughal court 'two … learned men with the Books of the Law [Old Testament] and especially with the Gospels'[6] so that they could engage with representatives of the many religions and belief systems present at court,[7] Islam and

---

christiano, indi sacerdote, e religioso della Compagnia, e chiamosli Francesco Enrichez, fatto consorte de' due ch'erano i principali, non perciò ch'egli fosse huomo per iscienza bastevole a così grande affare: peroche da quel primo dì che il lume della Fede gli entrò ad aprir gli occhi, e dargli vista, e conoscimento del vero, altro libro egli non usò di studiare più che il Crocifisso; ma perché in questo haveva profondamente appresa tanta sapienza di spirito, e messala in esercitio d'ogni più eccellente virtù, che tra per merito di santità, e per alcun tanto di lingua Persiana che possedeva, fu antiposto a molti altri di più riguardevoli qualità' ('But in the end, after a few days, the lot fell upon fathers Ridolfo Aquaviva and Antonio Monserrato, to whom it pleased to add a third, already by birth and sect of Moorish origin, then Christian, then priest, and member of the Company, called Francisco Enrichez, made consort of the two who were the principal missionaries, not because he was a man of sufficient science for so great a mission: for from that first day that the light of Faith entered him to open his eyes, and gave him sight, and knowledge of the truth, no other book he used to study more than the Crucifix; but because in this he had profoundly learned so much wisdom of spirit, and put it into the exercise of every most excellent virtue, that by merit of holiness, and by some amount of Persian language which he possessed, he was put before many others of more respectable qualities'), 'Ambasciadore inviato a Goa dal Re Achbar a condirne due Padri. Speranze che si havevano della sua conversione. Il P. Ridolfo eletto a quell'impresa', *Missione al Gran Mogor*, 12–21 (16).

6 The diplomatic missive requested 'due dei vostri uomini dotti con i libri della legge e soprattutto con i Vangeli, perché desidero sinceramente comprendere la loro perfezione … . Chiedo con grande urgenza che vengano con il mio ambasciator … e portino i loro libri. Dalla loro venuta trarrò la massima consolazione; mi saranno molto cari e li riceverò con ogni possibile onore. Appena sarò ben istruito nella legge e avrò compreso la sua perfezione, potranno tornare indietro a loro piacimento e li rimanderò da voi con grandi onori e ricompense adeguate'. Quoted from 'Firmano dell'imperatore Akbar ai Gesuiti del Collegio di San Paolo di Goa, dicembre 1578', in *Il trono di Moghul. La saga dei grandi imperatori dell'India* (Milan: Il Saggiatore, 2011), 177. The *farman* is also analysed in *Akbar e i Gesuiti*, 16–19.

7 Youri Martini lists the Hindus, Sikhs, Parsis, Buddhists, Jews, Christians, Jains, Sabians and

Hinduism in particular.[8] In this way, the *Padshah* aimed at accrediting himself as the benevolent ruler of an expanding empire that included numerous religions and belief systems. The mission with which Acquaviva and his companions were charged was a concrete response to this demand. Henriques, a native Farsi speaker (one of the official languages spoken at the Mughal court), had the key role of interpreter.[9]

Bartoli's volume narrates the unfolding of the mission, which, in Acquaviva's case, lasted until May 1583. From the Jesuits' perspective, it was a mission fueled by the hope that Akbar might convert to Christianity and had invited them to his court with this objective in mind. But in reality, the Jesuits' presence at court was part of Akbar's policy of *Sulḥ-i Kull*, or 'Universal Peace', a political and theological project to present the *Padshah* as a universal sovereign whose authority transcended the boundaries of any one religious law, and who therefore served as the indispensable mediator between the many different faiths of his subjects and of the larger world beyond its borders.[10] Thus, once Acquaviva

---

Charvaka; see *Akbar e i Gesuiti*, 27. The Charvaka, according to Amartya Sen, quoted by Martini, were an Indian philosophical school that denied the existence of God and that of the soul.

8 Rajeev Kinra, 'The Truth is out There (and also in Here): Taḥqīq as an Investigative Modality in Mughal Culture and Scholarship', *Journal of Early Modern History* 27, no. 4 (2023): 353–367.

9 For a critical analysis of documentation: M.S. Renick, 'Akbar's First Embassy to Goa. Its Diplomatic and Religious Aspects', *Indica* 7, no. 1 (1970): 33–47; John Correia-Afonso, 'Documents of the First Jesuit Mission from Goa to the Great Moghal', in *Actas do II Seminário Internacional de História Indo-Portuguesa*, ed. Luis de Albuquerque and Inacio Guerreiro (Lisbon: Instituto de Investigação Científica Tropical, Centro de Estudos de História e Cartografia Antiga, 1985), 29399; Adriano Mariotti, 'La prima missione dei Gesuiti alla corte di Akbar (1580–1583)', in *India tra Oriente e Occidente. L'apporto dei viaggiatori e missionari italiani nei secoli XVI–XVIII*, ed. Enrico Fasana and Giuseppe Sorge (Milan: Jaca Book, 1991), 75–100. See also the documentation on the report written in Latin by Antoni Monserrat, which remained unpublished until rediscovered and published in Calcutta by Henri Hosten SJ in the early twentieth century, then translated into English and recently into Catalan and Spanish: Antoni Monserrat, *Mongolicae legationis commentarius or The First Jesuit Mission to Akbar*, ed. Rev. H. Osten S.J., in *Memoirs of the Asiatic Society of Bengal* 3, no. 9 (1914): 513–704; *The Commentary of Father Monserrate, J.S., on His Journey to the Court of Akbar*, trans. J.S. Hoyland, commented by S.N. Banerjee (Oxford: Oxford University Press, 1922); *Embajador en la Corte Del Gran Mongol: Viajes de un jesuita catalán del siglo XVI por la India, Paquistán, Afganistán y El Himalaya*, trans. and ed. Josep Luís Alay e Ramón Sala (Barcelona: Editorial Milenio, 2006).

10 On the *Sulḥ-i Kull*, see Azfar Moin, ed., *Sulh-i Kul as an Oath of Peace: Mughal Political Theology in History, Theory, Comparison*, special issue of *Modern Asian Studies* 56, no. 3 (2022).

became fully aware that Akbar had no interest in embracing the Christian faith exclusively, he returned to Goa at the urging of his superiors.[11]

While introduced by an excursus on Acquaviva's education prior to his arrival in India, the work concludes with an account of the unfortunate martyrdom of Rodolfo and his four new confreres. Recklessly sent by their superiors to Orlim, a village on the Salcete Peninsula beyond Portuguese territorial control, to establish a church and missionary house there, the missionaries were slaughtered by the inhabitants. This happened in the violent context of the ongoing Portuguese conquest of the area, accompanied by forced conversions and by the destruction and desecration of temples and 'idols' (Hindu statues and images: *idoli* in the jargon of missionary sources), perpetrated also by the Jesuits themselves. Such actions had generated a deep aversion to Portuguese rule and to the fathers' Catholic proselytising.

The violent death of Acquaviva and his companions was recounted almost immediately after the fact by Alessandro Valignano S.J. (1539–1606) who in those very days of July 1583 was in the village of Cunculim, having accompanied to Cochim and Goa the Japanese legates who were on their way to the Roman curia on the long journey from Nagasaki to Lisbon (the well-known *Missio legatorum Iaponensium ad Romanam curiam*, 1582–1590). Valignano, who would have liked to accompany the Japanese legates as far as Rome, instead received, upon his arrival in Goa, an appointment as Provincial of the Indies, which required him to remain close to his mission sites in Asia.[12] Valignano's report of Rodolfo's violent death – intended primarily for the brethren of the Society in Europe and for the Jesuit general and Rodolfo's uncle, Claudio Acquaviva (1543–1615, general from 1581)[13] – was printed in the Italian vernacular as early as 1584 and subsequently reissued several times, including as part of the *littera annua* published in 1585.[14] Valignano recounts that in addition to Acquaviva

11 *Missione al Gran Mogor*: 'Detti, e fatti del medesimo Re, dirirtamente contrari alle speranze concepute [*sic*] doversi rendere Christiano', 58–70; 'Nuove speranze per le quali il P. Ridolfo si rimane tuttavia nel Mogor. Solitaria e santa vita che ivi menava', 71–77; 'Il Re Achbàr si fa istitutore d'una nuova Religione. Il P. Ridolfo il lascie, e torna a Goa', 77–84.

12 In an extensive bibliography, see Alessandro Valignano, *Dialogo sulla missione degli ambasciatori giapponesi alla curia romana e sulle cose osservate in Europa e durante tutto il viaggio*, ed. Marisa Di Russo, trans. Pia Assunta Airoldi (Florence: Olschki, 2016), 12.

13 For a broad summary of the cultural history of the long generalship of Claudio Acquaviva S.J., see Paolo Broggio, Francesca Cantù, Pierre-Antoine Fabre, eds, *I gesuiti al tempo di Claudio Acquaviva. Strategie politiche, religiose e culturali tra Cinque e Seicento* (Brescia: Morcelliana, 2007).

14 *Relatione della felice morte di cinque religiosi della Compagnia di Giesù, et di alcuni altri secolari ammazzati da Gentili per la fede nell'India Orientale l'anno 1583. Cavata da una del P. Allesandro Valignano Provinciale dell'india per il P. Generale della compagna di Giesù data in Goa alli 28 Dicembre dell'istesso anno* (Rome: Francesco Zanetti, 1584).

and four other Jesuits, several others were martyred in the incident: two young Brahmin (*Bramani*) Indian converts to Christianity remembered by their Portuguese given names, Domingo (who had been raised by the fathers and had explained the complex Hindu belief system to them) and Afonso, as well as two other adult Indians who had assisted the Jesuits by mediating with the Indian believers, known as Francisco Rodrigues and Paulo da Costa. In the massacre a Portuguese notary, Gonçalo Rodrigues, also died.[15]

Written a century after the events it describes and based on the letters sent by the three missionaries – primarily those of Rodolfo Acquaviva[16] – and, for the description of the events leading up to Rodolfo's martyrdom, on Valignano's *Relatione*,[17] the *Missione al Gran Mogor* was intellectually rooted in the desire for a historiographical and cultural systematisation of the Society's history, carried out by such preeminent figures as Jean Bolland S.J. (1596–1665),[18] Athanasius Kircher (1602–1680),[19] and Mathias Tanner (1630–1692),[20] to mention only the principal figures, in addition to Bartoli himself.

---

15 *Relatione della felice morte di cinque religiosi della Compagnia di Giesù*, [9].

16 J. Wicki, ed., *Documenta Indica*, vol. XII (Rome: Institutum Historicum Societatis Iesu, 1972), doc. 1, 3, 5, 6, 7, 8, 9 (1580); 50, 51 (1581); 106, 108, 109, 114 (1582); 127, 145 (1583).

17 P.A. Valignani S.I., provincialis, *Relatio de martyribus salsetanis*, Goa 8 Decembris 1583, *Documenta Indica*, doc. 145, 933–945.

18 As an example of their multi-faceted approach, see *Imago primi saeculi Societatis Iesu a Prouincia Flandro-Belgica euisdem Societatis repreasentata* (Antwerp: Plantiniana Balthasaris Moreti, 1640), compiled to commemorate the first centenary of the Society of Jesus and draw attention to the progress and achievements of the Society, eventually published by the Jesuit Press in Antwerp. The volume combined history, poetry and emblematics to illustrate the ideals, goals and achievements of the Society, interpreted as a living person, emulating in its actions and evolution over time the divine Person of Christ the Saviour.

19 In Kircher's extraordinarily extensive literary output, see his *China Monumentis, Qua Sacris quà Profanis, Nec non variis Naturæ & Artis Spectaculis, Aliarumque rerum memorabilium Argumentis Illustrata, auspiciis Leopold Primi, Roman, Imper. semper Augusti, Munificentißimi Mecœnatis* (Amsterdam: Janssonius, 1667). The six parts into which the work, known as *China illustrata*, is divided describe China and India, their languages (in particular, Chinese and Sanskrit), *monumenta* (e.g., the famous Xi'an stele erected by Nestorian Christians in 781 AD with inscriptions in Chinese and Syriac, or the Potala Palace in Lhasa, represented for the first time), the 'idolatries', their geography and history, mechanical arts and literature, especially of China. The work also presents an overview of the Society's missionary work in Asia.

20 Tarrell's encyclopedic history of the Society interpreted as a history of martyrs, recounted the deaths of numerous Jesuits at the hands of 'infidels and gentiles', with illustrations by artist Karel Škréta (1610–1675) and engraver Melchior Küsel (1626–1684): *Societas Jesu usque ad sanguinis et vitae profusionem militans, in Europa, Africa, Asia, et America, contra Gentiles mahometanos, judaeos haereticos, impios, pro Deo fide, ecclesia, pietate sive vita et mors eorumqui ex Societe Jesu in causa Fidei, & Virtutis propugnatae, violenta*

This chapter will briefly consider a few passages from the *Missione al Gran Mogor del P. Ridolfo Acquaviva* from his arrival in Goa and his subsequent stay at the court of the *Padshah* in Fatehpur Sikri (first together with Henriques and Montserrat, and then alone), focussing on the linguistic practices described, observed and experienced during the three years at the Mughal court, as described by Bartoli. This analysis is propaedeutic and at the same time instrumental in highlighting a paradigmatic transformation in linguistic knowledge, awareness and practices that took place within the framework of the Catholic missions from the mid-sixteenth century onwards and which, only a century later, at the time of Bartoli and Kircher, came to pervasively integrate European knowledge systems by fostering a pervasive attentiveness the existence of new languages.[21]

The *Mission to the Great Mogor* and Language Practices

As is customary in this kind of treatise, the *Missione al Gran Mogor* opens with a brief description of the Mughal Empire. The salient feature Bartoli highlights is the breadth and 'diversity' of the kingdom, also reflected by its linguistic pluralism: 'Né minor diversità, e confusione è nella moltitudine delle lingue. Il volgo, parla Moresco; i nobili del paese, Malavaro; la corte, Persiano bastardo; i savj, Arabo fino in servigio dell'Alcorano' ('Nor less diversity and confusion is in the multitude of languages. Commoners speak Persian; the nobles of the kingdom, Malavar; the court, spurious Persian; the scholars, Classical Arabic in the service of Alcoran').[22] The *Padshah* Akbar, 'the legitimate and seventh descendant of Tamerlane',[23] is described in encomiastic tones that highlight his intellec-

morte toto Orbe sublati sunt. Auctore R. Patre Mathia Tanner e Societate Jesu, ss. Theologiae Doctore (Prague: Typis Universitatis Carolo-Ferdinandeae per Joannem Nicolaum Hampel factorem, 1675).

21 See Angelo Cattaneo, *Tradurre il mondo. Le missioni, il portoghese e nuovi spazi di lingue connesse nella prima età moderna* (Rome: Bulzoni Editore, 2022).

22 'Brieve contezza del Regno del Gran Mogor. Qualità naturali e morali del Re Achabar', *Missione al Gran Mogor*, 7–12 (9).

23 'Hor fin dall'anno 1556 era l'imperio del Mogòr per legittima successione venuto alle mani d'Abdul Gelål Edim Achabàr, settimo erede del Tamerlano', *Missione al Gran Mogor*, 3. The reference is to Tīmūr Barlas or Tīmūr Lang, i.e., Timur 'the Lame', known in Europe as Tamerlane (1336–1405), a leader of Mongolian descent who conquered a large part of Central Asia, founding the Timurid empire between 1370 and 1405 following the disintegration of the Mongolian empire. See Michele Bernardini, *Tamerlano, Il conquistatore delle steppe che assoggetto l'Asia dando vita a una nuova civiltà* (Rome: Salerno Editore, 2022).

tual curiosity, especially regarding religions. He appreciated the confrontation between the exponents of Islam and Hinduism and brought them together to watch and listen to their disputes:[24]

> L'ingegno l'havea perspicacissimo, e ne dava segno la vivacità degli occhi che gli brillavano in fronte: ma di lettere per istudio, non havea né puro una lieve tintura, anzi né anche sapeva leggere il suo medesimo nome; e nondimeno vaghissimo era di tener molte ore innanzi a sé attizzati, e commessi in disputa i Mulassi, e i Bramani, quegli a difendere l'Alcorano, questi non so quali loro antiche scritture ...
>
> ([Akbar] had a most perspicacious wit, and the liveliness of the eyes that shone in his forehead gave sign of it: but of letters by study, he had not even a slight tincture, nay, he could not even read his own name; and nevertheless he had a great desire to keep the Mullahs (*Mulassi*)[25] and the Brahmins (*Bramani*) committed to dispute in front of him for many hours, the former to defend the Alcoran, the latter I know not which of their ancient scriptures ...).[26]

Bartoli then quickly summarizes Acquaviva's education before he was sent to India[27] and, once he landed in Goa, vividly recounts Rodolfo's sudden discovery of linguistic otherness and the specific forms of pedagogy and of cultural and linguistic translation implemented by the missionaries:

> Il P. Ridolfo ancor fresco della navigatione d'Europa, l'invitò a godere un giorno delle delitie di quell'isoletta; le quali furono, farlo incontrare a più di quattrocento fanciulli, stesi in lunga processione a due a due, tutti medesimamente atteggiati colle braccia recatesi in croce sul petto, e cantanti in lor lingua la *Dottrina christiana*, che l'Apostolo S. Francesco Saverio haveva composta in una semplice forma di versi, acconci a una tal'aria di musica, ond'ella era facile a ben comprenderfi, e dilettevole a recitarsi. Compiuto c'hebbero il lor canto, tutti a un medesimo tempo, in atto d'esecratione, sputarono (dissero) in faccia a gl'idoli: che così

24 'Ambasciadore inviato a Goa dal Re Achbar a condirne due Padri. Speranze che si havevano della sua conversione. Il P. Ridolfo eletto a quell'impresa', *Missione al Gran Mogor*, 12–21.

25 The Muslim religious scholars, in India, Iran and Turkic-speaking countries.

26 *Missione al Gran Mogor* (1714), 5.

27 'Ristretto della vita di P. Ridolfo fino alla partenza per l'India', *Missione al Gran Mogor* (1714), 22–33.

> erano ammaestrati: poi fattisi al P. Ridolfo, il ringratiarono in lor lingua, dell'esser egli venuto da un sì lontano mondo colà in prò delle anime loro. Hor l'effetto che una si bella mostra di christiana pietà cagionò in lui, fu un tenero lagrimare in quanto ella durò; e da quell'hora sentirli morto nel cuore ogni desiderio di studiare, e di sapere, se non solamente i linguaggi dell'India, per potersi gittare in mezzo de' barbari idolatri, e farne di così belli acquisti alla Fede.[28]
>
> (Fr. Ridolfo, still fresh from the voyage from Europe, was invited one day to enjoy the delights of that little island [Divar];[29] which allowed him to meet more than four hundred children, in a long procession two by two, all alike posed with their arms crossed on their chests, and singing in their own language the Christian Doctrine, which the Apostle St. Francis Xavier had composed in a simple form of verse, suited to such an air of music, so that it was easy to understand and delightful to recite. When they had completed their song, they all at the same time, in an act of execration, spat (they said) in the face of the idols: that they were thus instructed: then when they had turned to Father Ridolfo, they thanked him in their own language, for having come there from so distant a world, in favour of their souls. Hence the effect which so beautiful an exhibition of Christian piety caused in him was tender weeping as long as it lasted; and from that hour to kill his heart any desire to study and to know except only the languages of India, so that he might throw himself into the midst of the barbarous idolaters, and make such beautiful acquisitions to the Faith.)

This extended quotation is revealing of the complexity, pervasiveness and, in the case of India, imposing aspects implicit in the linguistic practices put in place after Francisco Xavier's arrival in Goa in 1542. The encounter with the children, lined up in a long procession, arms crossed on their chests, singing the Christian Doctrine in verse in one of the local languages, probably Konkhani or Tamil, transformed into a melody and taught to them by heart – an image of great emotional impact, which Bartoli, in the context of his culture, recounted in commendatory form – stirs in Rodolfo the desire to learn the languages of India. At the time of Bartoli this was explicitly perceived as an indispensable step in carrying out missionary work. Acquaviva's hopes of being trained and destined for the mission, studying languages in order to devote himself to

28 *Missione al Gran Mogor* (1714), 32.

29 The small island of Divar, in the vicinity of Goa Velha.

preaching were nevertheless frustrated in the first instance: Father Rodolfo was destined to teach philosophy in the Jesuit seminary.[30] Soon, however, he was chosen to coordinate the mission to Akbar's court.[31] As reported by Bartoli, language practices were of considerable importance, and the choice to send Henriques had depended on his knowledge of the Farsi language. The kind of discussions and debates in which the three Jesuits became involved with the Mullahs and Brahmins, initially undertaken through interpreters, revealed the difficulties inherent in the linguistic and cultural translation of Christianity. Therefore, Rodolfo committed himself to learning Persian firsthand with a Persian master, a practice he continued throughout his stay at the court:

> Volle che il P. Ridolfo si desse allo studio della favella Persiana, per poterlo udire ragionar della Fede con quella proprietà di parole, che l'interprete non esprimeva: ed egli, con istupor della Corte, in poco più di tre mesi n'hebbe la lingua e la scrittura sì bastevolmente spedita, che il Re gli mandò recare in Persiano una non piccola parte de gli Evangelj, e se l'udiva leggere da Habd-ulfathi, il più avvenente, il più amato de' suoi cavalieri, giovane d'acutissimo ingegno, uditore de' Padri in segreto, e in palese difenditor della Fede, tutto che di professione *Mulaso*, e in dignità fra elli come Prelato. Da cotal lectione si formò nella mente del Re un così nobile sentimento dell'essere, e del potere di Christo, ch'egli diceva, al certo cosa più che d'huomo essere stata in lui quella virtù, onde potè, come, e quando gli era in piacere, operar tanti e sì stupendi miracoli.[32]
>
> ([Akbar] wanted Fr. Ridolfo to devote himself to the study of the Persian language, that he might hear him reasoning about the Faith with that propriety of words which the interpreter did not express: and he, to the astonishment of the Court, in a little more than three months succeeded in learning the language and writing so expeditiously, that the King ordered him to translate into Persian not a small part of the Gospels, and had it read by Habd-ulfathil [the court dignitary Abu 'l-Fadl, assigned by Akbar

30 'Ma quanto a ciò, Iddio havea destinato condurlo per tale strada, che all'umano giudicio pareva terminar tutto altrove, che dove in fine ella riuscì. Perché inviati da' Superiori la miglior parte de compagni seco venuti d'Europa, chi ad uno, chi ad un altro regno dell'India, ad apprenderne il linguaggio, e disporsi al ministero della divina predicatione, egli solo fu posto allo sterile, e secco mestiere d'insegnare a' nostri giovani filosofia' *Missione al Gran Mogor* (1714), 34.

31 'Va da Goa alla Corte del Gran Mogor: Come ivi accolto dal re. Nulla ne accetta in dono. Convince in disputa i Mulassi', *Missione al Gran Mogor* (1714), 40–49.

32 *Missione al Gran Mogor* (1714), 48.

> to the Jesuits as an interlocutor], the most capable, the most beloved of his knights, a young man of keenest intellect, a confidant of the Fathers in secret, and in overt defender of the Faith, although he was by profession a Mullah, and in dignity among them as a Prelate. From these translations and interlocutions there was formed in the King's mind so noble a feeling of the being and power of Christ, that he said, at a certain time, that there had been in [Christ] that virtue, whereby he could perform so many and so stupendous miracles, anytime it pleased him.)

The untranslatability and incommunicability of faith discourses through interpreters, is a topos that we also find in medieval missionary treatises, for example in the *Itinerarium fratris Willielmi de Rubruquis de ordine fratrum Minorum, Galli, Anno gratia 1253 ad partes Orientales* by Franciscan Guillaume de Rubrouck (ca. 1210–ca. 1270), from 1248 to 1254 the protagonist of one of the first documented journeys to the Mongolian court of Möngke Khan (1209–1259) in Karakorum.[33] Brother Guillaume often complained about his lack of knowledge of local languages and the inadequacy or lack of interpreters, which often caused him to desist from attempting to preach, and even communicate, beyond daily needs:

> Super omnia autem grauabat me quod quando uolebam eis dicere aliquod uerbum edificationis, interpres meus dicebat: 'Non faciatis me predicare, quia nescio talia uerba dicere!' Et uerum dicebat. Ego enim percepi postea, quando incepi aliquantulum intelligere idioma, quod quando dicebam unum, ipse totum aliud dicebat, secundum quod ei occurrebat. Tunc uidens periculum loquendi per ipsum elegi magis tacere.[34]
>
> (I was especially vexed by the fact that whenever I wanted to do some preaching to them my interpreter would say, 'Do not make me preach,

33 English translation and contextualization: Peter Jackson and David Morgan, *The Mission of Friar William of Rubruck: His Journey to the Court of the Great Khan Möngke, 1253–1255*, Hakluyt Society, Second Series (Abingdon: Routledge, 1990). For Rubruck's biography and an extensive contextualization, see 'Introduction', 1–55. Italian edition and contextualisation: Guglielmo di Rubruck, *Viaggio in Mongolia-Itinerarium. Testo latino a fronte*, ed. Paolo Chiesa (Milan: Mondadori, 2011). For Rubruck's biography and his journey, see 'Introduzione', ix–lvi.

34 *Itinerarium* XIII.6, *Viaggio*, 64. (*The Mission of Friar William of Rubruck*, 108). For further analysis see Cattaneo, *Tradurre il mondo*, 17–33. On the Mongols and interfaith dialogue, see Joni Brack, 'Disenchanting Heaven: Interfaith Debate, Sacral Kingship, and Conversion to Islam in the Mongol Empire, 1260–1335', *Past and Present* 250, no. 1 (2021): 11–53.

> since I do not know how to express these things'. He was right. Later, when I acquired some little knowledge of the language [Mongolian], I noticed that when I said one thing, he would say something totally different, depending on what came into his head. After that I realised the danger of speaking through him and chose rather to say nothing.)

Returning to the *Missione al Gran Mogor*, it is important to highlight that Abu 'l-Fadl, of Muslim faith, a master of Persian and confidant of Acquaviva, at the invitation of Akbar and in interlocution with Rodolfo, translated the gospels into Persian; this was part of the activities of the *Maktab Khana*, a school of Persian translations founded by the *Padshah* in 1575 to translate major works from the cultures and religions present at the Mughal court, brought together through royal invitations, as had been the case with the Jesuits.[35]

Bartoli's account then continues with the famous description of Akbar who, without warning, visited the small chapel built in the court by the Jesuit fathers, in which two images had been displayed, arousing Akbar's admiration. One of the images (the second one is not described) was a reproduction of the famous Byzantine icon known as the 'Madonna of St. Luke', picturing the Virgin and the infant Jesus, preserved in the church of Santa Maria del Popolo in Rome. Akbar visited the chapel again on another occasion, accompanied by three of his sons, his brother-in-law and Abu 'l-Fadl and they all prostrated themselves with devotion before the two images.[36] These sentiments of apparent Christian devotion, however, were elements of Akbar's syncretic view of religions and rituals as part of a political project to rule an empire marked by a multiplicity of religions, philosophies and languages, kept together through an integrated military elite, centralised power, a culturally pluralistic cultural politics and both innovative and effective pan-Indian cultural and linguistic translation policies. The founding of the *Maktab Khana*, the school of translation into Persian, played an important role in Akbar's political vision, one not devoid of a utopian dimension: the *Padshah* thought that Persian could become the main idiom for pan-Indian communication, and that the translatability of religious truths, through the aegis of imperial patronage, was itself an expression of Akbar's 'Universal Peace'.[37] Bartoli eloquently describes this

35 *Akbar e i Gesuiti*, 47–48. See also Audrey Trushke, 'The Mughal Book of War: A Persian Translation of the Sanskrit Mahabharata', *Comparative Studies of South Asia, Africa and the Middle East*, 31, no. 2 (2011): 506–520.

36 *Akbar e i Gesuiti*, 49–50.

37 *Akbar e i Gesuiti*, 25. For a case study on the philosophy of translatability and its connection to Akbar's project of universal sovereignty, see Dan Sheffield, 'Exercises in Peace: Āẕar

syncretic understanding of religions, quoting his thought verbatim: '[secondo Akbar] Iddio è da onorare con ogni maniera d'adoratione, che comunque elle sieno di qualunque diversa religione, tutte a lui si convengono, e stanno bene.'[38] ('[According to Akbar] God is to be honoured with every manner of adoration, that although they may be of a different religion, they all agree with [God], and are good.') Akbar's syncretic attitude of mutual acceptance and peaceful coexistence among different religions extended to the construction of places of worship: just as Muslims and Brahmins had sumptuous mosques and pagodas available for their worships and rites, so too should Christians, in his understanding, have adequate churches and buildings available.[39]

This attitude, of welcoming but not adhering to Christianity, while frustrating the fathers' expectations for a possible and rapid conversion of the *Padshah*, was at the same time used as a pretext, Bartoli recounts, for an attempt by Muslim rivals to revolt against the political centralism constructed by Akbar. This, prompted by an alliance that brought together the Islamic kingdoms of Bengal, Kabul, the Afghan territories and Bihar, was organised by the ulama and headed, at least nominally, from Kabul by one of Akbar's half-brothers, Mirza Muhammad Hakim. The latter tried in vain to usurp the throne, engaging Akbar in violence on the battlefield witnessed by, among others, one of Acquaviva's companions, Antoni de Montserrat, who left the court of Fatehpur Sikri and accompanied the *Padshah* on his military campaigns.[40]

It is important to remember that, in the same years in which this first Jesuit mission was taking place in Fatehpur Sikri, the Mughal Empire and the Portuguese *Estado da Índia* faced each other militarily for the first time near the port cities of Diu and Daman. The conflict arose on the northwest coast of India, in relation to the navigation rights of Muslim ships crossing the Indian Ocean from Gujarat to the Arab ports of Jeddah, the Arabian Peninsula and the Persian Gulf, carrying Muslim pilgrims doing the *hajj*, the holy pilgrimage to Mecca.[41] Subsidised in part by the Mughal kingdom, the pilgrims' ships were targets of reprisals by the Portuguese, who imposed a sea transit tax on them by issuing passes (*cartazes*). However, the Mughal attack on Portuguese territory was repelled and contained by the Portuguese, becoming the basis of an agree-

Kayvānī Universalism and Comparison in the School of Doctrines', *Modern Asian Studies* 56 (2022): 959–992.

38 Quoted in *Akbar e i Gesuiti*, 49.

39 *Missione al Gran Mogor* (1714), 51.

40 *Akbar e i Gesuiti*, 28 and the extensive bibliography cited.

41 See Jorge Flores, *The Mughal Padshah. A Jesuit Treatise on Emperor Jahangir's Court and Household* (Leiden: Brill, 2015), 10–21.

ment to safeguard the pilgrims in exchange for a pact of non-belligerence by the Mughal ships.[42] It is in this conflictual context that Bartoli reports another relevant aspect of the linguistic practices in missionary contexts:

> Egli medesimo [Akbar] diè un de' suoi figliuoli al P. Monserrate, a fin che diligentemente gl'insegnasse favellare, e scrivere Portoghese; ben sapendo, come egli dise, che in fine sotto tal maestro l'havrebbe non meno Christiano nel cuore, che Portoghese nella lingua.[43]
>
> (He himself [Akbar] entrusted one of his sons to Fr. Monserrate, so that he would diligently teach him to speak and write Portuguese; knowing well, as he said, that in the end under such a master he would have [his son] no less Christian in heart than Portuguese in language.)

Following the interruption of the first mission in 1583, the Jesuit missionary presence was renewed in two more missions, starting in 1590, during Akbar's reign and then, following a sudden interruption in 1591, starting again in 1594 until 1614, for a stay that continued even after Akbar's death, during the reign of his successor Jahangir (1569–1627), the fourth *Padshah* of the Mughal realm, who reigned from 1605 until his death in 1627.[44]

Another episode reported by Bartoli occurred during the second Jesuit mission to Lahore in 1590, where the capital was moved from 1585, at the expense of Fatehpur Sikri.[45] As highlighted by Martini, 'the ruler asked the missionaries to start a school to teach to write and read Portuguese to the children of dignitaries and nobles, including probably a son of Akbar, and some members of the royal family'.[46] From the Jesuits' point of view, these practices were framed within an epistemological paradigm already outlined around 1540 by the Lusitanian chronicler and grammarian João de Barros (1496–1570), which will be discussed below, in the final part of this chapter. According to Barros, ultramar-

42 Michael N. Pearson, 'Portuguese India and The Mughals', *Proceedings of the Indian History Congress* 59 (1998): 407–426; Mika Natif, *Mughal Occidentalism. Artistic Encounters between Europe and Asia at the Courts of India, 1580–1630* (Leiden: Brill, 2018), 26–67.

43 *Missione al Gran Mogor* (1714), 51.

44 On these topics, see the comprehensive studies by Henri Didier, 'Les jésuites auprès du Grand Mogol Akbar (fin XVIe siècle): une mission chrétienne atypique auprès d'un souverain musulman atypique?', in *L'Islam des marges: mission chrétienne et espaces périphériques du monde musulman XVIème-XXème siècles*, ed. Bernard Heyberger and Rémy Madinier (Paris: Karthala, 2011), 17–43, and Flores, *The Mughal Padshah*.

45 John Correia-Afonso, 'The Second Jesuit Mission to Akbar', *Indica* 28 (1991), 73–93 (74–75).

46 *Akbar e i Gesuiti*, 59.

ine expansion should integrate the spread and imposition of the Portuguese language with proselytisation in Catholic Christianity, in the context of empire building and consolidation. From the Mughal perspective, on the other hand, the acquisition of the Portuguese language and of knowledge about the Catholic faith, including the translations of the Gospels, were instead rooted and framed in a project of cultural expansion of the Mughal empire in which, as noted above, a plurality of faiths, religions and languages should coexist under the sovereign's stewardship.

In the context of the second and third mission to the Persian court, beginning in 1590, the Jesuit fathers' efforts to learn and deepen their knowledge of the Persian language had developed and consolidated since Acquaviva's time, with the preparation of further translations of the Gospels, followed by a grammar of Persian and a manuscript trilingual Latin-Portuguese-Persian dictionary, titled *Rudimenta Linguae Persicae*. Although conventionally attributed to Jerónimo de Ezpeleta y Goñi (1549–1617),[47] the trilingual dictionary, now preserved at the School of Oriental and African Studies in London, was the outcome of at least two decades of linguistic interaction between the missionaries and their numerous Persian-speaking interlocutors at the Mughal court.[48]

Better known as Jerónimo Xavier, because of his distant kinship with St. Francisco Xavier, the Apostle of the Indies (1506–1552), Jerónimo de Ezpeleta y Goñi stayed in Lahore until 1614 and then for a year in Agra before returning to Goa. He was accompanied by the Portuguese Jesuit Manoel Pinheiro (1552–1619), who dealt mainly with the Indian population outside the Mughal court, and by the *irmão* Bento de Goís (c. 1562–1607), an excellent connoisseur of the Persian language, which he had learned when in Hormuz as a soldier in the Portuguese army. De Gois not only developed a solid relationship with the *Padshah* Akbar but also helped to organise a Mughal embassy to Goa between 1600 and 1601. After the embassy, in 1602, De Gois set out again, managing to reach China on foot, passing through Afghanistan and Tibet, proving empirically that the 'Cathay' described by European missionaries and merchants in the Middle Ages corresponded to the territories of China reached by Jesuit missionaries from 1580 onward.[49]

47 Henri Didier, 'Jérome Xavier, conseiller chrétien d'Akbar (1542–1605), roi aussi machiavélique que musulman?', in *Islam et christianisme. Éthique et politique*, ed. M.T. Urvoy (Paris: Éditions de Paris, 2010), 103–141.

48 SOAS London, Ms. 12198. See Flores, *The Mughal Padshah*, 43. Conventionally attributed to Xavier, this trilingual manuscript dictionary is more likely to be the collective effort of the missionaries and their Persian interlocutors, as Flores argues convincingly. The work is currently being studied by Flores and the Persianist Stefano Pellò.

49 *Akbar e i Gesuiti*, 84–88.

Missions and New Spaces of Connected Languages

The language practices described by Bartoli around 1650 (by order of citation: children reciting and singing the *Christian Doctrine* in Konkhani or Tamil in procession; Rudolph's intense desire to learn the languages of India; his learning of Persian at the Mughal court while working with court dignitary Abu 'l-Fadl to translate the Gospels into Persian; the theological disputes in Persian with the mullahs; the willingness of one of the *Padshah*'s sons to learn Portuguese; and the consolidation of the Jesuits' knowledge of Persian during the second and third mission to the court, beginning in 1590, following Acquaviva's death) reveal a full awareness of the importance of linguistic interaction in missionary practices. Three distinct synchronic and partly connected cultural paradigms explain or at least contextualise these practices.

As previously mentioned, a first paradigm combined Portuguese imperial expansion with language diffusion and Catholic proselytising. At the origin of this paradigm was the conceptualisation of the mission of the Empire and the Portuguese language articulated by João de Barros (1496–1570), former governor of the castle of São Jorge da Mina (1522–1525), the treasurer of Portuguese proto-colonial institutions such as the *Casa da Índia*, the *Casa da Mina* and the *Casa de Ceuta* (1525–1528) and feudal lord of the *Casa de Guinea* and the *Casa da Índia* (1533–1567). In 1539, he was also responsible for an important, albeit disastrous, colonisation plan for Brazil, which ended in a shipwreck. Alongside these official duties, Barros authored the *Décadas da Ásia*, a monumental chronicle that celebrated Portuguese navigations and conquests in Africa and Asia (mainly India), in three volumes published in 1552, 1553 and 1563, followed by a fourth, which was published posthumously in Madrid only in 1615. Just as importantly, Barros had the printer Luís Rodrigues publish, in Lisbon on 20 December 1539, the *Grammatica da língua Portuguesa com os mandamentos da Santa madre Igreja* (Grammar of the Portuguese Language with the Commandments of the Holy Mother Church) also known as *Cartilha com os preceitos e mandamentos da Santa madre Igreja, e co[m] os Misterios da missa e Responsoreos Della* (Primer with the Precepts and Commandments of Holy Mother Church and with the Mysteries of the Mass and its Responsories). The *Cartilha* served as an introduction to the *Grammatica da língua Portuguesa*, published a few days later, on 12 January 1540, with the addition of the *Diálogo em louvor da nossa linguagem* (Dialogue in Praise of our language).

In the *Cartilha*, João de Barros explained that the purpose of the grammar was to expound the rules and structure of the Portuguese language for foreigners, especially in the overseas territories, and for the 'children of these kingdoms' to learn to read and love it. Barros's conceptualization was rooted in the paradigm of the *lengua, compañiera del Imperio*, defined in the *Gramática de la lengua castellana* by humanist Antonio de Nebrija (1441–1522), printed in Salamanca in 1492. According to Nebrija, language was an indispensable factor in the identity of kingdoms and peoples, a theme already taken up in Portugal by Fernão de Oliveira (1507–1581) in the *Grammatica da lingoagem portuguesa*, printed in Lisbon in 1536. On the model of the Roman Empire, language could become an instrument of power in the process of expansion and conquest.[50] This model would later translate and expand, following Columbus's voyages for the crown of Castile, into the 'Castilianisation' of the newly conquered territories of *Nueva España*. Barros transforms Nebrija's paradigm by adding to the dyad 'empire-building' and 'imposition and spread of the Spanish language' the religious proselytising of Christianity, thus giving rise to the triad 'Empire, Language, Catholic Religion'. Thus, from the first half of the sixteenth century, the political and commercial expansion of the Portuguese empire was also thought of as a viaticum not only for the expansion of the Portuguese language but also of the Catholic faith.[51]

Within the framework of this 'imperial paradigm', Catholic missionaries working in the multilingual and multireligious spaces of the Iberian Empires and Royal Patronages (*Padroados*) from about 1530 onward enacted a fundamental cultural transformation that gave rise to a second major paradigm for linguistic interactions and practices: they described and translated numerous Amerindian, African and Asian languages, largely previously unknown to Europeans. Spanish and Portuguese were the predominant translation idioms, while Latin was used for grammatical descriptions. Simultaneously, in the American continent, the Iberian languages were imposed as part of the Spanish and Portuguese imperial expansion,[52] while along the extensive African, Persian and Asian coasts, Portuguese joined Arabic, Persian, Chinese and Malayalam as one of the *lingua francas* of maritime and port communities.[53] In

50 Sheldon Pollock has highlighted how the link between language and empire is a feature of Western societies that originated with the Roman Empire and is absent in the Sanskrit tradition of South Asia. See Sheldon Pollock, *The Language of the Gods in the World of Men, Sanskrit, Culture, and Power in Premodern India* (Berkeley: University of California Press, 2006), 270–271, 274–276.

51 Simone Celani, *Alle origini della grammaticografia portoghese* (Rome: Edizioni Nuova Cultura, 2012) 61–79.

52 Walter Mignolo, *The Darker Side of the Renaissance: Literacy, Territoriality, and Colonization* (Ann Arbor: University of Michigan Press, [1995] 2003).

53 Hugo Cardoso, 'O português em contacto na Ásia e no Pacífico', in *Manual de linguística portuguesa*, ed. Ana Maria Martins and Ernestina Carrilho (Berlin: De Gruyter, 2016), 68–97.

both the West and East Indies, Catholic missionaries, in asymmetrical power contexts and yet interacting with local communities, drafted numerous pioneering catechisms and doctrines in Amerindian, Asian and African languages, as well as bilingual or trilingual lexicographical works, accompanied by bilingual grammars, and translations (especially into Asian languages) of classical and Christian authors.[54]

Beginning around 1530, as part of the Franciscan and Dominican missions on the American continent, Spanish – first and predominantly conceived as the 'companion' language of power, according to Antonio de Nebrija's conceptualisation – also became the idiom of description and translation for many Mesoamerican and Andean languages and, from around 1565, also for some of the languages of the Philippines. The phenomenon has already been extensively investigated by others, to whose studies we refer.[55] In the plural and polycentric context of Portuguese expansion and the Catholic missions connected to the Empire and Royal patronage, Portuguese, in addition to being *língua companheira do Império*, was over the course of two centuries also the main medium of translation, along with Latin and, in a more limited form, Spanish and Italian, for numerous languages then unknown to Europeans.

A geographically ordered list, which makes no claim to exhaustiveness, includes in Asia: Tamil and Malayalam (Dravidian languages spoken in southeastern India, Kerala and along the northeastern coasts of Sri Lanka); Konkani, Marathi and Bengali (Indo-Aryan languages of the Indian subcontinent spoken in Goa, Maharashtra and Bengal); Chinese, and Japanese. In Brazil: the *língua geral paulista*, *língua geral* and *língua geral amazônica* (Tupi and Tupi-guaraní languages). In Africa: Kimbundu, Kikongo and Sena (Bantu languages spoken in sub-Saharan Africa and, as a consequence of the slave trade, also in Brazil).[56] To get an idea of the extent of this phenomenon, consider that in the same year,

54 Otto Zwartjes, *Portuguese Missionary Grammars in Asia, Africa and Brazil, 1550–1800* (Amsterdam-Philadelphia: John Benjamins, 2011); Simone Celani, *Alle origini della grammaticografia portoghese* (Rome: Edizioni Nuova Cultura, 2012); Antje Flüchter and Rouven Wirbser, eds, *Translating Catechisms, Translating Cultures. The Expansion of Catholicism in the Early Modern World* (Leiden: Brill, 2017).

55 Otto Zwartjes et al, eds, *Missionary Linguistics I–VI*, (Amsterdam: John Benjamins, 2004–2021).

56 Maria Leonor Buescu, *A galáxia das línguas na época da expansão*, (Lisbon: Comissão Nacional para as Comemorações dos Descobrimentos Portugueses, 1992); Maria Helena Mira Mateus, *Caminhos do português: exposição comemorativa do ano europeu das línguas: catálogo* (Lisbon: Biblioteca Nacional, 2001); Telmo Verdelho and Joao Paulo Silvestre, eds, *Dicionarística Portuguesa. Inventariação e Estudos do Património Lexicográfico* (Aveiro: Universidade de Aveiro, 2007); Carlos Assunção and Gonçalo Fernandes, 'Contributos da Língua Portuguesa para as Descrições das Línguas do Oriente', in *Pontes entre Lusofonia*

1595, the Jesuit press published both a grammar of Tupi-Guarani – printed in Coimbra but prepared and edited in the mission of São Paulo by Spanish Jesuit José de Anchieta (1534–1597) in collaboration with local students and mission assistants[57] – and, on the opposite side of the world in Amakusa (Kyūshū, Japan), a trilingual Latin-Portuguese-Japanese dictionary, based on Ambrose Calepino's Latin dictionary, comprising 27,000 entries for each of the three languages.[58]

On the basis of the documentary surveys carried out within the scholarly framework of Missionary Linguistics – a branch of the history of linguistics that since the beginning of the twenty-first century has begun to review and aggregate in a systematic form the linguistic documentation produced in Catholic missionary contexts – starting from the first decades of the sixteenth century, it is possible to draw up a cultural geography of the languages described and first translated into European languages through Spanish and Portuguese.[59]

---

e Francofonia: A aliança da 'latinoesfera', ed. Isabelle Oliveira (Porto: Edições Afrontamento, 2017), 61–79. For a systematic analysis of these linguistic interactions see Cattaneo, *Tradurre il mondo*, 35–132.

57 *Arte de Crammatica da lingoa mais usada na costa do Brasil. Feyta pelo padre Ioseph de Anchieta da Cõpanhia de IESV* (Coimbra: António Mariz, 1595; ed. Julio Platzmann, Leipzig: B.G. Teubner, 1874, 1876; Rio de Janeiro: Biblioteca Nacional, 1933; São Paulo, 1946; Salvador: Universidade Federal da Bahia, 1980; São Paulo: Loyola, 1990).

58 *Dictionarium Latino Lusitanicum, ac Iaponicum ex Abrosii Calepini volumine depromptum: in quo omissis nominibus proprijs tam locorum, quàm hominum, ac quibusdam alijs minùs usitatis, omnes vocabulorum significationes, ele antioresque dicendi modi opponuntur: in usum, & gratiam Iaponicae iuventutis, quae Latino idiomati operam nauat, nec non Europeorum, qui Iaponicum sermonem addiscunt.* In Amacusa in Collegio Iaponico Societatis Iesu cum facultate Superiorum 1595.

59 See the following volumes in the *Missionary Linguistics* series in particular: Otto Zwartjes and Even Hovdhaugen, eds, *Missionary Linguistics – Lingüística misionera: Selected papers from the First International Conference on Missionary Linguistics*, Oslo, 13–16 March, 2003 (Amsterdam-Philadelphia: John Benjamins, 2004); Otto Zwartjes and Cristina Altman, eds, *Missionary Lingustics II – Linguistica misionera II: Orthography and Phonology: selected papers from the Second International Conference on Missionary Linguistics*, São Paulo, 10–13 March 2004 (Amsterdam-Philadelphia: John Benjamins, 2005); Otto Zwartjes, Ramón Arzápalo Marín and Thomas C. Smith-Stark, eds, *Missionary Linguistics IV – Lingüística misionera IV: Lexicography: selected papers from the fifth International Conference on Missionary Linguistics*, Mérida, Yucatán, 14–17 March 2007 (Amsterdam-Philadelphia: John Benjamins, 2009); Otto Zwartjes, Klaus Zimmermann and Martina Schrader-Kniffki, eds, *Missionary Linguistics V – Lingüística Misionera V: Translation theories and practices: Selected papers from the Seventh International Conference on Missionary linguistics*, Bremen, 28 February–2 March 2012 (Amsterdam-Philadelphia: John Benjamins, 2014); Otto Zwartjes and Paolo de Troia, eds, *Missionary Linguistics VI. Missionary Linguistics in Asia. Selected papers from the Tenth International Conference on Missionary Linguistics*, Rome,

Scrutiny of the sources highlights the emergence of new multilingual spaces, communities and identities in the heterogeneous contexts of missions related to the territories of the Portuguese Empire in the period 1500–1700. These documents include primers (*cartilhas* or *cartinhas*) for learning primary literacy skills (the Latin alphabet and Portuguese syllables) combined with an abridged catechism, Christian doctrines (*doctrinas christaãs*) that summarise and outline the precepts of the Christian faith, sometimes in dialogue form, as well as grammars (*artes da língua*) and dictionaries (*vocabularios*). Collectively, they offer a fundamental documentary framework and an original contribution to an analysis of the global cultural history of early modernity, as well as its periodisation and spatialisation.

In sub-Saharan Africa, particularly in the Kingdoms of Congo and Angola, the first linguistic interactions began in the mid-1500s and were consolidated by the end of the century through the initiative of the Portuguese crown, in collaboration with Franciscans and the first local converts.[60] But it was especially with the founding of the first Jesuit missions in Asia, beginning in 1542, with the arrival of Francisco Xavier S.J. (1506–1552) in India, Japan and China, and then around 1570, with the arrival of Father *Visitador* Alessandro Valignano S.J. (1539–1606), that practices of linguistic and cultural learning, description, codification and translation were implemented in a systematic form in India, Japan, China, Tonkin (in today's Vietnam), Brazil and vast coastal regions of sub-Saharan West Africa. In all these areas Portuguese was the main language of contact and translation.[61]

The two paradigms that I have so far briefly outlined explain and contextualise, from a missionary documentary point of view, the linguistic practices that Bartoli, having access to and studying the documentation preserved in the Jesuit *Arquivum historicum*, highlighted in the *Missione al Gran Mogor*, within the general framework of his *Istoria della Compagnia di Giesù*. However, the history of the Jesuit mission to the Mughal court of Akbar and later his heir Jahangir, allows us to highlight a third paradigm, synchronous to the two already analysed, which helps to explain the linguistic and missionary practices of Acquaviva and his confreres in the Mughal kingdom. While the desire of the Jesuits Rodolfo Acquaviva, Antoni de Monserrat and later Jerónimo Xavier

21–24 March 2018 (Amsterdam-Philadelphia: John Benjamins, 2021).

60 Diogo Ramada Curto, *Cultura imperial e projetos coloniais (séculos XV a XVIII)* (Campinas: Unicamp, 2009); Diogo Ramada Curto, *Imperial Culture and Colonial Projects. The Portuguese-Speaking World from the Fifteenth to the Eighteenth Centuries*. Translated from the Portuguese by Alison Aiken (New York: Berghahn Books, 2020).

61 Cattaneo, *Tradurre il mondo*, 35–87.

and Manoel Pinheiro to learn Persian was rooted in the practices put in place in Goa, the study and promotion of learning Persian in the colleges and seminary of Rachol, from the beginning of the mission in India – which focused on the need to learn local languages in order to carry out missionary activity – was also rooted in the translation policy pursued through the founding of a specific institution, the *Maktab Khana*; this was the Persian translation school, equipped with a scriptorium, founded and supported by Akbar at the court in Fatehpur Sikri, around 1574.[62]

The *Padshah* had assembled a very rich intellectual environment at court, which included scholars and clerics from a wide range of faiths, schools and languages, endeavouring to have a library of works translated into Persian, the language of the court, the military elite, the nobility, the government and the administration; it included, at least ideally, the sacred books of humankind, the major works of scientific and philosophical knowledge originally in Arabic, Sanskrit, Greek, Latin and other European languages.[63] Very significant was, for example, the Persian editing of an abridged version of the *Mahābhārata* (The Great [story] of the Bhārata), a Hindu epic poem originally in Sanskrit, titled in the Persian translation *Razmnama* (Book of War), completed around 1598–1599; this was followed later by the Persian translation of other Hindu epics, such as the *Rāmāyaṇa* (Path of Yama), along with encyclopaedias in Arabic, the Qur'ān and, as anticipated, with the arrival of the first Jesuit mission, particularly Rodolfo Acquaviva, also the annotated Gospels.[64]

In this connection, it is worth mentioning that in the same years, beginning in 1584, the *Tipografia medicea orientale* (Medici Oriental Press) was in operation, first in Rome and then in Florence, founded by Cardinal Ferdinando de' Medici (1549–1609, from 1587 third Grand Duke of Tuscany) and directed by the orientalist Giambattista Raimondi (1536–1614), who devoted himself to printing in non-Latin characters. He published, among other things, editions of the Gospels in Arabic and Arabic-Latin, grammars of the Arabic and Syriac languages and classical Arabic and Persian texts of philosophy, geography, and mathematics. The venture was not as commercially successful as

---

62 On the exchange between the Persian-speaking Islamic elite of the Mughal Empire and Sanskrit scholars under emperors Akbar, Jahangir (1605–1627), and Shah Jahan (1628–1658), when Sanskrit epics and historical texts, were translated into Persian, see Audrey Trushke, *Culture of Encounters: Sanskrit at the Mughal Court* (New York: Columbia University Press, 2016).

63 *Akbar e i Gesuiti*, 24.

64 Muzaffar Alam, 'The Pursuit of Persian: Language in Mughal Politics', *Modern Asian Studies* 32, no. 2 (1998): 317–349; M. Athar Ali, 'Translations of Sanskrit Works at Akbar's Court', *Social Scientist* 20, nos. 9/10 (1992): 38–45.

hoped, yet it marked a crucial moment in the history of European Orientalism and early attempts to systematise the Arabic, Syriac and Persian languages.[65]

It was at the interface of these three synchronic political and cultural paradigms – which collectively shaped the forms, practices, and purposes of cultural and linguistic interactions in the decades between 1570 and 1620 – that Rodolfo Acquaviva, his confreres who continued missions to the Mughal court and their far more numerous local interlocutors at the Mughal court (such as the dignitary Abu 'l-Fadl, the mullahs and *ulama*, the Brahmins, even the *Padshah* Akbar and Jahangir themselves) were involved in complex interactions; while these were sometimes marked by rivalries and carried out with heterogeneous and even divergent goals, they were not without collaborative, shared aspects as well. The learning of the Persian language, in priority, but also Portuguese – both of which had risen to the status of *lingua franca*, in the neighbouring territories of the Mughal empire and the *Estado da Índia* respectively – was rooted in the belief shared by Akbar and the missionaries (particularly the Jesuits) that languages and knowledge were mutually translatable among themselves and that it was possible to implement cultural translations even within very heterogeneous, if not incompatible, ideological frameworks and aims.[66] The Jesuits interpreted these cultural and linguistic translations, which also included forms of adaptation to local cultures (accommodation), as a function of exclusive conversion and adherence to Christianity.

By contrast, the *Padshah* configured them as a strategy within the framework of a new political project of sovereignty based on the belief that the plurality of faiths and knowledge, conveyed by heterogeneous languages, practices and rituals in the broad context of the Mughal empire, could be harmonised by the shared and sharable search for truth, thought of and experienced as an *unicum*, which could be drawn upon through dialogues, theological disputes and cultural and linguistic translations.

65 See Sara Fani and Margherita Farina, eds, *Le vie delle Lettere. La Tipografia Medicea tra Roma e l'Oriente*. Catalogo della mostra: Firenze, Biblioteca Medicea Laurenziana, 26 ottobre 2012–2022 giugno 2013 (Florence: Mandragora, 2013), and Eckhard Leuschner and Gerhard Wolf, eds, *The Medici Oriental Press: Knowledge and Cultural Transfer around 1600* (Florence: Olschki, 2022). Both volumes study the origins, publishing and business activities, and cultural environment of the *Tipografia*, also surveying its documentary base and the manuscript and printed volumes produced.

66 Richard Eaton, *India in the Persianate Age, 1000–1765* (Berkeley: University of California Press, 2019), 25.

In these far-reaching politico-cultural processes, the Persian language, through the school of translations founded by Akbar, was being rethought as a *lingua franca* and an instrument of pan-Indian cultural and political cohesion. Portuguese, which had already become one of the languages of communication in the Indian Ocean through the Lusitanian empire and its mercantile networks, especially Jesuit missionaries, had been transformed instead by missionaries into a language of translation to support missions, for a wide range of language families unknown in Europe until the mid-sixteenth century.

Even within the multilingual framework of medieval Christianity,[67] scrutiny of mediaeval mercantile and missionary sources reveals that travel, trade and religious proselytising in Asia, India and China at the time of the Mongol Chinese Yuan Empire (c. 1274–1368) and the so-called *Pax Mongolica* had not resulted in a coherent, consolidated and transmissible knowledge of the many languages spoken beyond the 'multilingual space of the Bible' (that is, beyond the spaces of the cultures and peoples of the Mediterranean, Persia and the Arabian Peninsula). The idioms that mediaeval merchants and missionaries, mainly Franciscans, could not recognise or translate through their own personal expertise or that of the interpreters they employed, were referred to with a residual taxonomy as 'self-contained languages', in the sense of autonomous, unrelated, untranslatable languages (*lingue per sé* in Italian vernacular or *proprium idioma et propria littera* in Latin). This attitude changed substantially as language practices and cultural interaction promoted and implemented in the context of missions, especially in the Portuguese context, became established and consolidated.

Rodolfo Acquaviva's cultural biography, traced by Bartoli on the basis of his letters, the reports of his confreres involved in the missions to Akbar's court and, more generally, with deliberate anachronism, of the *documenta indica* preserved in the Jesuit *Arquivum*, is also permeated by these dynamics and sharply documents a very broad missionary, cultural and existential perspective, beyond martyrdom.

Over the course of about 150 years, important linguistic interactions developed within the framework of Catholic missions, which, from the perspective of cultural history, are an extremely interesting laboratory of cross-cultural, political, linguistic and religious interactions.[68] In this relatively short period,

67 Albrecht Classen, 'Multilingualism in the Middle Ages and the Early Modern Age: The Literary-Historical Evidence', *Neophilologus* 97 (2013): 131–145.

68 See Joan-Pau Rubiés, 'From Idolatry to Religions: The Missionary Discourses on Hinduism and Buddhism and the Invention of Monotheistic Confucianism, 1550–1700', *Journal of Early Modern History* 24 (2020): 499–536; Joan-Pau Rubiés, 'Comparing Cultures in the

small communities of European Catholic missionaries experienced very heterogeneous forms of cultural interactions with much larger local political, mercantile, military and religious communities. The codification, teaching and mutual learning of previously disconnected languages were also part of and went hand in hand with the founding and organisation of pedagogical institutions, such as Jesuit colleges and seminaries in India, Brazil, Japan and China, but also in the case of the Mughal Empire, the *Maktab Khana*, the Persian translation school founded by Akbar in 1575. The conspicuous linguistic interactions we have analysed were a constitutive part of ambivalent power relations in which cooperation, acceptance, confrontation and rejection, dialogue and imposition, understanding and misunderstanding were in constant dialectic.

Missionaries such as Rodolfo Acquaviva and his confreres sought to address and negotiate the radical asymmetries of power with their local interlocutors, adopting and exploring strategies of adaptation that, over the course of several decades, ensured the implementation of relevant processes of mutual learning. Far from being an aprioristic philosophical stance of (presumed and wholly anachronistic) respect for cultural differences or otherness or, on the contrary – a mere strategy to penetrate and conquer the (otherwise deemed lost) consciousnesses of gentiles – the Jesuits' 'adaptation' to local contexts of interaction in early modernity, the result of radical power asymmetry, were prototypical forms of mutual cultural translation.[69]

Reading and analysing the epistolary correspondences and reports composed as early as the first thirty years of the Jesuits' missionary activity in India, Japan, Brazil and China, reveal that the inculturation and acculturation of Christianity, later systematised by Alessandro Valignano in works composed after his first stay in Japan between 1579 and 1582 – such as, the *Advertimentos e Avisos acerca dos costumes e catangues de Jappão* (1581),[70] the *Sumario de las cosas de Japón* (1583)[71] and the *Historia del principio y progresso de la Compañía*

Early Modern World. Hierarchies, Genealogies and the Idea of European Modernity', in *Regimes of Comparatism: Frameworks of Comparison in History, Religion and Anthropology*, ed. Renaud Gagné, Simon Goldhill and Geoffrey Lloyd (Leiden and Boston: Brill, 2018), 116–176.

69 For a critical assessment of the *status quaestionis*, see Andrés I. Prieto, 'The Perils of Accommodation: Jesuit Missionary Strategies in the Early Modern World', *Journal of Jesuit Studies* 4 (2017): 395–414.

70 Alessandro Valignano, *Il Cerimoniale per i missionari del Giappone. "Advertimentos e Avisos acerca dos Costumes e Catangues de Jappão"*, ed. Josef Franz Schütte S.J. (Rome: Edizioni di Storia e Letteratura, 1946).

71 Alessandro Valignano, *Sumario de las cosas de Japon (1583). Adiciones del Sumario del Japan (1592)*, ed. José Luis Alvarez-Taladriz (Tokyo: Sophia University, 1954).

de Jesús en las Indias orientales (1583)[72] – depended largely on the practices that some members had tried to implement since the early years of the arrival of the missionaries of the Society of Jesus. Rodolfo Acquaviva was among them. The imbalance of power and demographics not only did not prevent the development of relevant cultural interactions but, on the contrary, it stimulated some Jesuits towards creative strategies and practices of cultural communication and translation which later became consolidated in the broader, global framework of early modern Catholic missions.

Acknowledgements

The research for this essay was conducted as part of the project funded by the Italian Ministry of Universities and Research (MUR): PRIN 2022 – 'Mapping and Translating Spaces, Cultures and Languages. Experiences from the Missions Connected to the Portuguese Empire (1540–1700)' – Ref.: 20222SY2K7 – CUP: B53D23001120006.

72 Alessandro Valignano, *Historia del principio y progresso de la Compañía de Jesús en las Indias orientales (1542–1564)*, ed. Josef Wicki (Rome: Jesuit Historical Institute, [1944] 2011).

CHAPTER 8

Language, Translation and Mobility across the Mediterranean and Beyond

Giancarlo Casale

As the attentive reader will have noticed, the lion's share of contributions to this volume have dealt with mobility and translation within a specific part of the world, or more precisely, across a specific body of water: the Mediterranean. This is quite by design, as the original conversation from which this collection of essays emerged posited the Mediterranean as an especially fertile site for the circulation of both individuals and texts. The reasons are numerous and, in some sense, obvious. First, as a self-contained maritime space, which has maintained over the millennia its own distinct cultural and ecological fingerprint even while connecting the earth's three most diverse continents, the Middle Sea has long been celebrated as both a conduit and a crucible of cultural intermixing.[1] Inevitably, an essential component of this intermixing has always involved linguistic encounters, the Mediterranean to this day being home to an extraordinary cacophony of languages hailing from at least four entirely un-related language families: Semitic, Afro-Asiatic, Indo-European and Turkic.[2] Just as important, many of these languages boast rich and in some cases extremely ancient textual traditions, phonetic writing itself being a Mediterranean invention named, most appropriately, for the first ancient maritime nation to trade across its full geographic span.[3]

As a result, it is no accident that the Mediterranean region has, over the course of its history, been the site of all three of the best documented and, arguably, most consequential translation movements of the pre-modern world: First, the Abbasid translation of texts from Greek, Syriac, Sanskrit and Pahlavi into Arabic between the eighth and the tenth centuries CE, then the medieval Iberian translations of Arabic texts into Latin from the eleventh to the fourteenth centuries, and finally, the Humanist translation movement from Greek

1 See Peregrine Horden and Nicholass Purcell, *The Corrupting Sea: A Study of Mediterranean History* (London: Wiley-Blackwell, 2000).

2 Piera Molinelli, ed., *Languges and Identity in Multilingual Mediterranean Settings: Challenges for Historical Sociolinguistics* (Berlin: Degruyter, 2017).

3 Stephen R. Fischer, *The Story of Writing* (London; Reaktion, 2021).

 | DOI:10.1163/9789004731806_010

(as well as Arabic, Syriac and Hebrew) into Latin and various Western vernaculars that began in the fifteenth century.[4] Thereafter, the Mediterranean continued to serve as either the birthplace or the functional hub of the most important intellectual and technological drivers of early modern translation as well, including (to name just the most obvious) the birth of the news industry, the institutionalization of modern diplomacy and the foundation of the *Propaganda fide*.[5]

And yet, it is also true that by celebrating the Mediterranean's uniquely rich history of linguistic encounter one also runs the risk of running aground on some of its most contested historiographical shoals. Of these, perhaps the rockiest of all involves the perennial debate over the essential 'unity' or 'diversity' of the sea's Muslim and Christian halves, neatly described by Eric Dursteler as the debate between 'battlefield' and 'bazaar' – in other words, over whether the Middle Sea is at heart a zone of cultural hybridity and symbiotic exchange, or instead of perpetual conflict between opposing civilizations.[6] Here, perhaps unsurprisingly, some very fierce arguments about translation – or lack thereof – across the Mediterranean have frequently served as a proxy for this larger, politically charged debate. Bernard Lewis, for example, several years before famously coining the term 'Clash of Civilisations' in a 1990 article on Islam and global politics, presaged his views in *The Muslim Discovery of Europe*, a much more scholarly book-length investigation of pre-modern translation and travel literature.[7] Therein, he documented an ostensible lack of interest among pre-modern Muslims in translating texts of European origin – a lack which he

4 On the Abbasids, Dimitri Gutas, *Greek Thought, Arabic Culture: The Graeco-Arabic Translation Movement in Baghdad and Early Abbasid Society* (London: Routledge, 1998); on the mediaeval Mediterranean, Karla Mallette, *Lives of the Great Languages: Arabic and Latin in the Medieval Mediterranean* (Chicago: University of Chicago Press, 2021); on the Renaissance, Dag Nikolaus Hasse, *Success and Suppression: Arabic Sciences and Philosophy in the Renaissance* (Cambridge MA: Harvard University Press, 2016).

5 On Humanism and translation, see James Hankins, 'Translation practice in the Renaissance: The case of Leonardo Bruni', in *Humanism and Platonism in the Italian Renaissance*, ed. J. Hankinsn vol. I (Rome: Edizioni di storia e letteratura, 2003), 177–192; On information and diplomacy, Filippo de Vivo, *Information and Communication in Venice: Rethinking Early Modern Politics* (Oxford: Oxford University Press, 2009); on print culture, Brenda M. Hosington, 'Translation and Print Culture in Early Modern Europe', *Renaissance Studies* 29, no. 1 (2015): 5–18; on the Jesuits, Peter Burke, 'The Jesuits and the Art of Translation in Early Modern Europe', in *The Jesuits II Cultures, Sciences, and the Arts, 1540–1773*, ed. John W. O'Malley (Toronto: University of Toronto Press, 2006), 24–32.

6 Eric Dursteler, 'On Renaissance Bazaars and Battlefields: Recent Scholarship on Mediterranean Cultural Contacts', *Journal of Early Modern History* 15, no. 5 (2011): 413–434.

7 Bernard Lewis, *The Muslim Discovery of Europe* (New York: W.W. Norton, 1982).

described, through language that surely reflected his own twentieth-century experience more than that of his pre-modern subjects, as an unassailable 'iron curtain' around Islam that began to be pierced only with the arrival of Western colonial modernity. In substance, this argument was intended as a response to the views of Lewis's intellectual nemesis, Edward Said, who had proposed only a few years earlier, in *Orientalism*, that Western translations of Islamic texts should be interpreted as an essential expression of Europe's intellectual domination of the Muslim world – and the lack of an equivalent 'Occidentalism' among Muslims the inevitable result of its condition of political and military subjugation to the West.[8]

Despite these opposing positions, Louis and Said did agree with one another on at least one basic point: that Mediterranean translation was essentially a one-way street, an activity that Westerners engaged in when describing the Orient, but not the reverse. But this is a view that, in more recent decades, has been challenged, qualified, and complicated by a great deal of new research, including on practices of communication (such as the elusive but quintessentially Mediterranean *Lingua Franca*) and on previously unstudied sources in Arabic, Turkish and other Islamic languages.[9] In particular, a quite prominent focus of recent scholarship has involved dragomans, professional interpreters of the early modern Mediterranean who can be situated only precariously on one side or the other of the supposed line dividing 'East and West'. In the Ottoman Empire, these were predominantly Christians, but also Ottoman subjects, and typically learned translation from their families as a kind of inherited artisanal trade. Most of them found employment with Western consuls and diplomats, while being habitually distrusted by these states for their 'in-between-ness'. And yet, as recently argued by Natalie Rothman, they were also creators in the first instance of the emerging colonialist discourse of contrasting 'civilisations' – a self-serving discourse from their perspective, which consolidated a general perception of the dragomans' own indispensability as technical experts in bridging an otherwise impenetrable cultural divide between Orient and Occident.[10]

8 Edward Said, *Orientalism* (New York: Knopf Doubleday, 1979).

9 Eric Dursteler, 'Speaking in Tongues: Multilingualism and Multi-Cultural Communication in the Early Modern Mediterranean', *Past and Present* 217, no. 1 (2012): 47–77; Karla Mallette, 'Lingua Franca', in, eds., *A Companion to Mediterranean History*, ed. Sharon Kinoshita and Peregrine Horden (Oxford: Wiley, 2014), 330–334; Nabil Matar, *Europe through Arab Eyes* (New York: Columbia University Press, 2009).

10 Natalie Rothman, *Dragoman Renaissance: Diplomatic Interpreters and the Routes of Orientalism* (Ithaca: Cornell University Press, 2021). See also Stefan Hanss, *Narrating the Dragoman's Self in the Veneto-Ottoman Balkans* (London: Routledge, 2023).

Dragomans thus serve as the most emblematic example of the new 'post-Orientalist' approach to Mediterranean studies that has come to dominate the field in the decades since Lewis and Said, characterized by a focus on hybridity, connectivity and the constructed nature of difference. Very recently, however, there are signs that a shift in the prevailing winds of scholarship is once again under way – seen, for example, in the growing appetite for '(dis)connected history' among early modernists – and as every Mediterranean navigator knows, a shift in the wind necessarily demands a corresponding correction of course.[11] For students of translation, one possibility for such a correction involves a return to the enduring differences in cultures of translation between the Ottoman and Western European regions of the Mediterranean, but with a new sensitivity to the contingencies involved, and a novel set of questions about what these differences implied.

Of these differences, perhaps the most superficially obvious is the Ottoman world's perplexing lack of enthusiasm for print technology until well into the eighteenth century – a vexing and much debated point that very recent scholarship has recharacterised as comparatively less important than previously thought.[12] Meanwhile, a much less frequently acknowledged but potentially critical point of social divergence involves the comparative mobility of individuals (including, of course, translators) between the Ottoman and Christian halves of the Mediterranean. Generally speaking, most European Christians of the early modern centuries – like most Western tourists today – assumed that the territory of the Ottoman empire was on the whole accessible to them, in the sense that travel there was routinely allowed except under extraordinary circumstances (such as the outbreak of war). Tellingly, one of the most important prerogatives that European consular authorities in the Ottoman empire negotiated for their subjects was not the right to travel and trade in Ottoman territory, as this was regularly allowed in any case, but rather the right to remain there *without* automatically becoming residents – and therefore taxpayers.[13] Importantly, however, the reverse was generally not true. That is, with the partial exception of Livorno, Venice and a handful of other port cities, it was essentially illegal for a Muslim to be physically present in most parts of Mediterranean

11 See Zoltan Bidermann, '(Dis)connected History and the Multiple Narratives of Global Early Modernity', *Modern Philology*, 119, no. 1 (2001): 13–32.

12 Nir Shafir, *The Order and Disorder of Communication: Pamphlets and Polemics in the Seventeenth-Century Ottoman Empire* (Palo Alto: Stanford University Press, 2024).

13 Maurits H. Boogert, *The capitulations and the Ottoman legal system: qadis, consuls, and beraths in the 18th century* (Leiden: Brill, 2005).

Europe, except under special conditions and with special legal provisions.[14] In practice, this basic difference in the legal regimes of travel meant that there were exponentially more Western Christians who freely travelled in the Ottoman empire than the reverse – a reality with obvious implications for language acquisition, and of course translation.

One such implication is that, for Muslims, the *primary* legal condition allowing them to be present in Christian Europe was enslavement. In this respect, although slavery was a shared Mediterranean institution, mutually inflicted by Muslims against Christians and vice versa with roughly equivalent methods and frequency, the social role of slavery and the life prospects of the enslaved were nevertheless considerably different in Islamic and Christian lands.[15] For most Muslims enslaved by Christians, the outlook was bleak indeed: those not fortunate enough to be ransomed or otherwise returned home could expect, with a few notable exceptions, a life of bondage or of permanent social marginalisation.[16] This was largely true even for those who chose, either out of conviction or through duress, to forsake Islam and embrace the religion of their captors, since, in much of Mediterranean Europe, converts from Islam (and in the case of Iberia, even the *descendants* of converts) were viewed with deep suspicion.[17] By contrast, in the Muslim Mediterranean, enslavement was – however counterintuitively from our modern perspective – a key mechanism of social mobility. This is not to deny, of course, that the actual conditions of slavery could be quite brutal. But as a legal system which, in practice, permanently bonded master and slave (even after emancipation) through reciprocal ties of obligation, Islamic slavery was widely understood as a form of fictive kinship whose desired final outcome, through incorporation into the master's household, was conversion and social integration.[18] In the Ottoman empire, nowhere was this more true than from the standpoint of the Ottoman state itself, to the extent that *kuls*, or state slaves, were accorded a special status as a kind of imperial elite as members of the sultan's own household, and were in consequence systematically favoured over free-born Muslims for appointment

14 For a recent discussion of this problem, Eloy Martìn Corales, *Muslims in Spain, 1492–1814: Living and Negotiating in the Land of the Infidel*, trans. Consuelo Lópoez-Morillas (Leiden: Brill, 2021).

15 Daniel Hershenzon, *The Captive Sea: Slavery, Communication and Commerce in Early Modern Spain and the Mediterranean* (Philadelphia: University of Pennsylvania Press, 2018).

16 Salvatore Bono, *Schiavi Musulmani nell'Italia Moderna* (Napoli: Edizioni scientifiche italiane, 1999).

17 See Bartolomé Bennassar and Lucile Bennassar, *Les Chrétiens d'Allah* (Paris: Perrin, 2006).

18 Charles Wilkins, 'Slavery and Household Formation in Ottoman Aleppo, 1640–1700', *Journal of the Economic and Social History of the Orient* 56, no. 3 (2013): 345–391.

to the upper echelons of the Ottoman military administration.[19] As a result, for much of the early modern period, *most* members of the Ottoman governing elite (including the highest-ranking generals, governors, and viziers) were individuals who had begun their careers in service to the Sultan as Christian slaves.[20]

Inevitably, such vastly divergent social realities had a direct impact on the socio-demographic profile and potential life trajectory of a typical translator. In the Christian Mediterranean, we know of a handful of examples from the entire early modern period of Muslim captives who, through a combination of their own efforts and the patronage of their masters, became literate enough in a Western language and socially connected enough to their Christian captors to engage in intellectual activities that included translation. Here, the celebrated case of Hassan Muhammad al-Wazzan (a.k.a. Leo Africanus) serves as an illustrative example – attracting scholarly attention precisely because, in a European context, his profile as a convert-translator is exotic and unusual.[21] Note, however, that in the context of the Ottoman Mediterranean, his trajectory as a captive turned convert/translator would have instead represented a typical, even unremarkable case. In fact, for the entire period of Ottoman history prior to the nineteenth century, there are but a handful of examples of translators from European languages into Turkish or Arabic who were *not* either slaves or renegades or converts to Islam from one of the Ottoman empire's local communities of Christians or Jews – a rule fully confirmed by the Ottoman translators highlighted in the present volume of essays (notably in the chapters by Durmaz and Tezcan). In other words, without slavery and conversion, it is unlikely that there would have been much of *any* formal translation from Western languages into Turkish or Arabic during the early modern period. Free-born Ottoman Muslims who engaged in such activity were exceedingly rare, and their absence is particularly striking among the highest-educated members of Ottoman society, the graduates of elite imperial *madrasas*.[22]

19 Dror Ze'evi, 'My Slave, My Son, My Lord: Slavery, Family and State in the Islamic Middle East', in *Slave Elites in the Middle East and Africa*, ed. Toru Miura and John Edward Philips (London: Kegan Paul, 2000), 71–80.

20 Tobias Grafe, *The Sultan's Renegades: Christian-European Converts to Islam and the Making of the Ottoman Elite, 1575–1610* (Oxford: Oxford University Press, 2017).

21 Natalie Zemon Davis, *Trickster Travels: A Sixteenth-Century Muslim between Worlds* (New York: Hill and Wang, 2007).

22 Note that one of the few native Muslim translators, Osman of Timişoara (whose translations from German notably included the first Turkish-language history of the Holy Roman Empire), had himself been enslaved in Christian Europe for over a decade. See Osman

Just why this should be so remains something of a mystery, still not fully explained by scholarship. Objectively speaking, it is difficult to imagine a more deeply multi-lingual education than that of an imperial Ottoman *madrasa*, which expected graduates to acquire a deep knowledge of all three of the Muslim Mediterranean's main textual languages: Arabic, Persian and Ottoman Turkish.[23] These languages are, moreover, profoundly different from one another: as distinct, grammatically speaking, as English, Hungarian and Hebrew, and therefore at least as demanding to master as the Humanistic trio of Latin, Greek and French, a gold-standard of sorts for erudite language study in the early modern West. Why, then, did this Ottoman focus on rigorous language study not organically lead, at least for some madrasa graduates, to an interest in Western languages, in the way that Humanistic education seems to have done much more readily for the study of the Islamic languages among early modern Europeans?[24]

At least one partial answer to this question, still quite understudied by scholars, involves an aspect of written language that is peculiar to the Mediterranean: a close association between religion and alphabets, to the extent that the ultimate marker of one's membership in a religious community was not the language one spoke, but rather the alphabet one wrote it in. This is a phenomenon perhaps most easily recognized with respect to Judaism: Judeo-Arabic and Ladino, the two most widely spoken tongues among Jews of the Ottoman Mediterranean, are essentially Arabic and Castilian Spanish written in Hebrew characters (and therefore chronically mistaken for Hebrew by those unable to read them).[25] A similar situation held for Moriscos, the community of Iberian Muslims first forcibly converted and then expelled from their homeland, some of whom wrote in Arabic while others in Aljamiado, a form of Castilian composed in Arabic script (the language, among other things, of the source texts for Joseph Morgan's *Mahometism Fully Explained*, studied in this volume by Ann Thomson).

This attachment to script was by no means limited to Muslim and Jewish communities, instead reflecting a more general pattern that, across the Otto-

---

of Timişoara, *Prisoner of the Infidels: The Memoir of an Ottoman Muslim in Seventeenth-Century Europe*, ed. and trans. Giancarlo Casale (Berkeley: University of California Press, 2021).

23 Christine Woodhead, 'Ottoman languages', in *The Ottoman World*, ed. Christine Woodhead (London: Taylor and Francis, 2012), 143–158.

24 On the latter, see Andrew Bevilacqua, *The Republic of Arabic Letters: Islam and the European Enlightenment* (Cambridge, MA: Harvard University Press, 2020).

25 See Geoffrey Khan, 'Judeo-Arabic', and David Bunis, 'Judezmo (Ladino)', *Handbook of Jewish Languages*, ed. Lily Khan and Aaron Rubin (Leiden: Brill, 2017), 22–63, 366–451.

man world, also included virtually every denomination of Christians. Thus, in Anatolia, Turkish-speaking Orthodox or Armenian Christians adapted the Greek and Armenian alphabets to Turkish, in the Balkans, Catholic speakers of south Slavic wrote in Latin characters while Orthodox wrote in Cyrillic, and so forth. In addition, since the use of a different alphabet rendered a text illegible to those of another community, this was normally enough for a tongue to be considered a different language entirely – thus necessitating a 'translation' from one to the other. Interestingly, this kind of translation remains today an important scholarly activity in modern Turkey, a consequence of the language reform of 1928 whereby the country, as a part of its 'conversion' from a Muslim to a secular nationalist state, adopted a Latin alphabet, rendering the entire written record of Ottoman Turkish unintelligible to future generations unless translated.[26]

Without more in-depth research, we are still quite far from understanding precisely what this intimate connection between faith, identity and alphabets might have meant for the region's cultures of translation. That said, it seems at least reasonable to assume a strong connection between the act of translating from one alphabet to another – or even of knowing how to read a text in another alphabet – and the act of religious conversion. Thus, in a social context such as that of the Ottoman empire, in which conversion from other religions to Islam was commonplace – but conversion from Islam to another religion was a capital offense! – there may have been a strong social predilection to leave to converts the task of translating texts from Western languages (or, more precisely, from languages written in non-Islamic alphabets). Conversely, if the same sensibility did not hold sway in the Christian West, this may have to do, in part, with the comparatively much lower social standing of converts, but likely also with Christian Europe's long history of linguistic engagement with Greek, and, to a certain extent, also with biblical Hebrew, which normalised the practice of studying non-Latin alphabets in a way that was eventually transferred also to the languages of Islam.[27]

If the Mediterranean's long history of 'alphabetic communities' did have the potential to shape translations in this way, then at least one place where its influence might be observed is in the comparatively much higher emphasis placed by early modern Muslims on translation *between* Islamic languages. As mentioned above, a proper Ottoman *madrasa* education was, in theory, based

26 Wesley Lummus, 'Turkic Latinization within a Trans-national Perspective', (PhD diss., University of Minnesota, 2021).

27 See for example Laurence de Looze, *The Letter and the Cosmos: How the Alphabet Has Shaped the Western View of the World* (Toronto: University of Toronto Press, 2016), 67–82.

on mastery of the so-called *elsine-i selāṣe* or 'three languages', namely Arabic, Persian and Turkish. In practice, however, such mastery was attainable by only a tiny scholarly elite, for the most part trained in a handful of privileged institutions. Outside of this exclusive group, numbering perhaps no more than a few thousand people at any given time, full literacy in even two of these three languages was quite rare. In the Arab lands, knowledge of Persian or Turkish was quite infrequent even among elite scholars (indeed, there is a strong case to be made that the very idea of 'the three languages' was consolidated to deny Arabophobe scholars full membership in the empire's intellectual elite).[28] And elsewhere, there were many Muslims who were native speakers of other, entirely unwritten languages (such as Kurdish or Albanian), and for whom literacy of any kind meant mastering a foreign tongue – but rarely more than one. Even for Turkish speakers, the bar for literacy was raised quite high by the elaboration of an elite imperial form of Ottoman Turkish radically different from any spoken version of the language, and therefore barely comprehensible to 'native' speakers without considerable study.[29]

As a result, one way to understand Ottoman expansion is as a process of translation between the empire's centre and its many distinct communities of literate Muslims, as the state's physical growth took place alongside the consolidation of imperial institutions that were both increasingly centralised and increasingly associated with the Ottoman Turkish language.[30] This preference for Turkish was confirmed only gradually: through to the end of the fifteenth century, the Ottoman imperial chancery regularly issued documents not only in Persian and Arabic but also in Greek and occasionally even Italian.[31] Chronicles and courtly literature were also composed in all of these languages and, if anyone at the time had thought to make a prediction, they most probably would have assumed that the Ottomans would follow a path similar to their later imperial rivals to the east, Safavid and the Mughals – two dynasties who,

28 Baki Tezcan, 'The Law School of Mehmed II in the last quarter of the sixteenth century: A glass ceiling for the less connected Ottoman *ulema*', in *Ottoman War and Peace: Studies in Honor of Virginia H. Aksan*, ed. Frank Castiglione, Ethan L. Menchinger and Veysel Şimşek (Leiden: Brill, 219), 237–282.

29 Kristof D'Hulster, *Ottoman Poets and Poetics in the Sixteenth Century: Introducing Asik Çelebi's and Latifi's Biographical Dictionaries* (Edinburgh: University of Edinburgh Press, 2025), 1–42.

30 Selim Kuru, 'The Making of a Literary Tradition', in *The Cambridge History of Turkey: Volume 2, The Ottoman Empire as a World Power, 1453–1603*, ed. Suraiya Faroqhi and Kate Fleet (Cambridge, 2013), 548–592.

31 See for instance Julian Raby, 'Mehmet the Conqueror's Greek Scriptorium', *Dumbarton Oaks Papers* 37 (1983): 15–34.

like the Ottomans, were also originally Turkish-speaking, but who never tried to impose this as the main language of state or of high culture in their realms. Instead, Ottoman Turkish began to affirm itself in the early sixteenth century, in large measure through the consolidation of a centralised Ottoman system of elite *madrasas*, whose Turkish-speaking graduates increasingly dominated Imperial institutions.[32] Their rise, moreover, closely coincided with the Ottomans' rapid expansion into the Persianate east and, even more dramatically, the Arab south after the conquest of Mamluk Syria and Egypt in 1516–1517. The latter, in particular, resulted in an unprecedented cultural encounter between two intellectual worlds, as the sudden reality of conquest created new pathways, and a new necessity, for scholars, jurists, bureaucrats and other members of the literate classes from each side to come to terms with the other.[33]

As in contemporary Western Europe, this combination of mobility, bureaucratic centralisation and vernacular language proved a powerful engine for translation. Precise figures are lacking, but over the course of the sixteenth and seventeenth centuries there were certainly hundreds and, more likely, thousands of titles in every imaginable genre translated between the 'three languages' of Ottoman Islam – a figure which, in all likelihood, represents the most extensive translation movement in history to take place in the absence of print technology. The vast majority of these texts remains unstudied and, in many cases, even uncatalogued by modern scholars, meaning that the most basic contours of this intellectual movement are only beginning to be understood.[34] What is clear, however, is that the range of works translated was as wide as the Ottoman world itself, running the gambit from primers and popular literature to technical treatises, monumental histories and peripatetic philosophy. Impressionistically, it also seems to be the case that there were more translations from Arabic and Persian into Turkish rather than the reverse – an intriguing parallel to the (perceived) imbalance, discussed earlier, of translation between Islamic and European languages in favour of the latter. And, predictably, in many cases these translated texts also included original interven-

32 Abdurrahman Atçıl, *Scholars and Sultans in the Early Modern Ottoman Empire* (Cambridge: Cambridge University Press, 2018).

33 Helen Pfeifer, *Empire of Salons: Conquest and Community in the Early Modern Ottoman Lands* (Princeton: Princeton University Press, 2022).

34 Some preliminary considerations are to be found in Philip Bockholt, Hülya Çelik, Ludwig Paul and Ani Sargsyan, eds, *Elsine-i Şelāşe: A Cultural Analysis of Transmission and Translation in the Ottoman Empire* (Göttingen: Verlag, 2025). See also Gottfried Hagen, 'Translations and Translators in a Multilingual Society: A Case Study of Persian-Ottoman Translations, Late Fifteenth to Early Seventeenth Century', *Eurasian Studies* 2/1 (2003), 95–134.

tions by their translators, which frequently combined antiquarianism, philology, exoticism and learned commentary in a manner strikingly similar to the writings of contemporary European humanists who travelled in the Ottoman Orient.[35] Recent research has even shown that, beginning in the seventeenth century, the linguistic tools developed by these Ottoman translators (including intertextual glosses, primers, vocabularies and so forth) were also circulated in Europe and proved foundationally important for the first professionally trained European translators of Oriental languages.[36]

A final consideration regarding Ottoman translation involves some very recent research into the philosophical and theological meanings of translation in early modern Islam, and its resulting relationship to politics. Such considerations could begin with the widely held Muslim doctrine that the Qur'an is the literal word of God, and therefore untranslatable as a sacred text.[37] In the realm of governance, one possible consequence of this doctrine is to insist that the laws of the state, directly mandated by God, must be based on a literalist and fixed interpretation of sacred scripture – thereby leaving virtually no autonomous space for the sovereign as a legislator. But this was only one possible position, another being the contrasting belief that the literal text of the Qur'an was a superficial veil that hid more profound divine truths.[38] And while there could be many possible paths to gain access to these hidden truths, among the most radical was through the comparative study of religions – achieved, in part, through the translation of sacred texts between faiths.

A particularly eloquent articulation of this theory of 'sacred translatability', recently studied by Dan Sheffield, involves the followers of Azar Kayvan, a sixteenth-century Iranian mystic who held that all languages, and by extension all religions, were derived from a common source and were thus translatable. In a posthumously compiled text that preserves some of his writings, this doctrine is illustrated by comparing equivalent technical terms drawn from various

35 For example, Giancarlo Casale, 'Time and the Other: Ottoman Encounters with Egypt's Ancient Past', *1516: The Year that Changed the Middle East and the World*, ed. Adulrahim Abu-Husayn (Beirut, 2021), 150–175.

36 Paul Babinsky, 'Ottoman Philology and the Origins of Persian Studies in Western Europe: The Gulistan's Orientalist Readers', *Lias* 42, no. 2 (2020): 233–315.

37 M. Brett Wilson, *Translating the Qur'an in an Age of Nationalism: Print Culture and Modern Islam in Turkey* (Oxford: Oxford University Press, 2014).

38 These two positions are generally associated with the opposing epistemologies of *taqlid* and *taḥqīq*. See Mathew Melvin-Koushki, '*Taḥqīq* vs. *Taqlid* in the Renaissances of Western Early Modernity', *Philological Encounters* 3 (2018): 193–249. Also Giancarlo Casale, 'Introduction: Cultures of Taḥqīq between the Mongols, the Mughals, and the Mediterranean', *Journal of Early Modern History* 27, no. 6 (2023): 273–287.

religious traditions (eg. 'pundit' in Sanskrit, 'mutakallim' in Arabic, 'mūbad' in Persian), the essential intelligibility of these terms across confessions leading to the conclusion that 'the difference occurs in the signifier and not in the signified, just like the words for "water": *āb* [Persian], *mā* [Arabic], sū [Turkish], and *pānī* [Hindi]'.[39] What is particularly suggestive about such 'sacred translatability', however, is that when transported into an imperial context, it could be deployed to support a theory of universal sovereignty, in a way that allowed a sovereign to free himself from the shackles of sacred law precisely through his support for the translation of sacred texts. Thus, in the 1570s, during a period exactly contemporary with Azar Kayvan, intellectuals at the court of the Mughal Emperor Akbar 'the Great' (r. 1556–1605) began to espouse a doctrine of 'Universal Peace' (*Ṣulḥ-i Kull*) that placed the emperor above Islamic law as the universal arbiter between the many diverse religious communities of the Mughal realm (of which Muslims were a distinct minority).[40] As such, one of the most tangible expressions of this doctrine in the realm of cultural patronage was for Akbar to commission officially sanctioned Persian translations of the sacred Sanskrit epics, the *Muhabharata* and the *Ramayana*.[41] In the same spirit, Akbar also sent a letter to Philip II of Spain, inviting him to send to his court 'learned men with the Books of the Law [Old Testament] and especially the Gospels' – thereby setting the stage, as discussed by Angelo Cattaneo in his contribution to the present volume, for Acquaviva's 'mission to the Great Mogul' and the subsequent account by Daniello Bartoli.

In comparison to the Mughals, Ottoman sultans are generally assumed to have been more invested in 'Sunni legalism' to legitimise their rule, and are therefore rarely associated with promoting this brand of 'sacred translatability'. But there were notable exceptions. In the fifteenth century, the reign of the foundational Ottoman sultan Mehmed the Conqueror has been productively compared to that of Akbar in terms of political ideology and, by no means coincidentally, Sultan Mehmed also supported the translation of an impressively eclectic body of religious texts, ranging from the Biblical *Book of Daniel* (from Syriac), the *Chaldean Oracles* (from Greek), a gloss on the *Nicene Creed* (from Latin), and even *On the dogmas of the Zoroastrians and Platonists*, a treatise by

39 Daniel J. Sheffield, 'Exercises in peace: Āẕar Kayvānī, Universalism and Comparison in the School of Doctrines', *Modern Asian Studies* 56/1 (2022), 979.

40 A. Azfar Moin, '*Sulh-i Kull* as an Oath of Peace: Mughal Political Theology in History, Theory, and Comparison', *Modern Asian Studies* 56, no. 2 (2022), 721–748.

41 Audrey Truschke, 'The Mughal Book of War: A Persian Translation of the Sanskrit *Mahabharata*', *Comparative Studies of South Asia, Africa and the Middle East*, 31, no. 2 (2011): 506–520.

the neo-Pagan philosopher George Gemistos Pletho.[42] Building on Mehmed's reign as a precedent, future research may well provide similar explanations for later moments in Ottoman history when official support for translations became comparatively more intense (such as the late-sixteenth-century reign of Sultan Murad III) or, conversely, help to better explain periods when translations – particularly from Western languages into Turkish and Arabic – proved unexpectedly rare. In his contribution to this volume, Baki Tezcan already points us in this general direction in discussing the seventeenth-century Polish-Lithuanian convert Ali Bey's translation of the suggestively titled Latin textbook *The Gate of Languages Unlocked, or the Seedbed of All the Languages and Sciences*, by John Amos Comenius. As Tezcan shows, there are reasons to think that Ali Bey's choice of this text may have been linked to a transcendental philosophy based on Ibn Arabi's 'Unity of Being' (*Waḥdat al-wujūd*), according to which all religions were emanations of a single divine truth. Intriguingly, Ali Bey is also described as belonging to a circle of contemporary like-minded Muslim intellectuals known as 'Christ lovers' (*hûb-Mesîhî*) – a group including prominent figures like Ebubekir of Damascus and Hüseyin Hezarfen who enjoyed close connections to the imperial court.

These Muslim 'Christ-lovers', pondering the essential translatablility of faiths in Ottoman Istanbul, lived only a generation or two before Joseph Morgan, who in *Mahometism Fully Explained* (as shown in Ann Thomson's contribution to this volume) used his own version of 'sacred translatability' to convince his English readers that the Muslim prophet, too, 'taught the worship of the real, the only, and the self-existing God, tho' after his own Manner'. Seen from this perspective, the cultural distance between Ottoman Istanbul and early modern London seems quite bridgeable indeed. But to what extent can this apparent convergence be attributed to actual opportunities for direct dialogue and, in consequence, shared ideas of translation? Rather than providing a single, definitive answer to such questions, the many contributions to this volume are akin to rhumb lines criss-crossing a well-worn portolan chart, each indicating a possible course across the sea still waiting to be charted.

---

42 Giancarlo Casale, 'Mehmed the Conqueror between *Sulh-i Kull* and *Prisca Theologia*', *Modern Asian Studies* 56, no. 2 (2022): 840–869.

CHAPTER 9

Situating Early Modern Translators: A Mediterranean Epilogue

Stefan Hanß

First comes space in the history of the early modern Mediterranean, Fernand Braudel taught us, and so it is also in the study of early modern translation.[1] As dictionaries of the time show, space was key to early modern understandings of the concepts and practices of translation, across languages. The Italian term 'to translate' (*traslatare*), the *Vocabolario degli Accademici della Crusca* states in 1623, meant first and foremost 'to transport [*trasportare*] [something] from one place to another place'. The dictionary stressed further that the word had acquired a second(ary) meaning only in more recent times: 'To reduce the writings and the compositions of one language into another one is today also called "to translate" [*tradurre*]', in particular if the translation was made into a vernacular language (*volgarizzare*, or *traslatare*). 'Translation' (*traslazione*), hence, was characterised by both space and language: *il traslatare, trasportamento*.[2] In Ottoman Turkish, too, 'to translate' was widely termed *tercüme* (ترجمه) but could be also rendered as *çevirmek* (چویرمك) or *naql* (نقلك) at times, which held the additional meanings of 'transfer' and 'transport'.[3] In the early modern Mediterranean, concepts of translation were thus deeply entangled with notions of space and mobility, and it is exactly during these centuries that the meaning of textual translation gains momentum and gradual prominence.[4]

Such semantics correlated with the ways early modern translation was practiced, as we see when turning our attention to translating practitioners as well as those commissioning translations. In 1617, for instance, the Venetian diplomatic resident in Istanbul Almorò Nani requested his in-house profes-

1 Fernand Braudel, *The Mediterranean and the Mediterranean World in the Age of Philip II*, vol. 1 (Berkeley: University of California Press, 1995), 25–167.

2 *Vocabolario degli Accademici della Crusca* (…) (Venice: Sarzina, 1623), 886, 892, 893.

3 J. W Redhouse, *Redhouse's Turkish Dictionary* (…), 2nd revised edition (London: Quaritch, 1880), 341–342.

4 Umberto Eco, *Experiences in Translation*, translated by Alastair McEwen (Toronto: University of Toronto, 2001), 74.

 | DOI:10.1163/9789004731806_011

sional translators (*dragomans*) to 'copy them [the letters] in Turkish … and to transport them [*trasportarle*] also into our language'.[5] In an audience with the Ottoman grand mufti in the same year, the Venetian diplomat's head dragoman Marc'antonio Borisi spoke about his work as a space of negotiation: a path 'to walk, … step by step'.[6] Interpreters working in Venetian diplomatic service, in fact, were expected to have a wide skillset in cultural, linguistic and spatial mobility, or *pratiche del passaggio* as Venetian authorities used to call it.[7] In practice as much as in theory, translation was therefore tied to movement.

This volume moves across words and worlds to capture the dynamism of translation and the vibrant world of translators in the early modern Mediterranean and beyond. The contributors' focus on mobility and translation, and the cultural and mobile worlds of early modern translators more precisely, is thus a pivotal contribution to our understanding of translations, and a very welcome and very much needed intervention in the fields of the history of translation, and translation studies more broadly. This volume uncovers the mani-facetted lives of translators, and the extent to which biographic events and developments, and experiences of spatial mobility in particular, impacted translation projects and agendas. What makes this volume unique is its emphasis on the necessity of more inclusive stories of translators across the early modern world and its insistence on the significance of the southern and eastern shores of the Middle Sea, as well as the southeastern European Ottoman world, for wider cultures of translation at that time. The Ottoman realm, we learn, experienced a watershed 'translation movement' that went largely unnoticed in the wider scholarship on early modern translations. Also the volume's emphasis on Mediterranean 'alphabetic communities' generates new insights into the complex relationship between language variety and cultural belonging in the early modern Mediterranean. The contributors' shift in focus towards space, thus, creates an awareness of significant gaps in the literature, and it presents case studies to address such.

This book renders visible those who are usually prone to invisibility: translators. Far too often, researchers of translations discuss early modern texts without a wider contextualisation of those who authored them. As re-written or twice-written texts, translations introduce an additional layer of complexity

5 Archivio di Stato di Venezia, Venice, Senato, Dispacci, Dispacci degli ambasciatori e residenti, Costantinopoli, filza 84, no. 8, November 9, 1617, fol. 111^{r}.

6 Ibid., fol. 109^{v}.

7 Stefan Hanß, *Narrating the Dragoman's Self in the Veneto-Ottoman Balkans, c. 1550–1650* (Abingdon: Routledge, 2023), 70–104.

and ambiguity to the notion of authorship.[8] This involves decisions about how to articulate, position, write and relate words, and decisions about the degree to which translators make such agency visible, or not. Countering what Lawrence Venuti has called the 'translator's invisibility', the volume introduces us to early modern translators as full-fledged historical protagonists.[9] It is the complexities of early modern translators' lives, one might say, that shape the lives of translations. This focus includes reflections on the ambivalent and often difficult circumstances in which language proficiency mattered and under which translations have been produced, including the everyday hardships of translators and the difficulties of making a living. In line with recent research that highlights the social constraints of early modern linguistic diversity, several chapters explicitly highlight the importance of slavery, captivity and other forms of involuntary or forced mobility, and immobility.[10] Such experiences were interwoven into the social fabric of the early modern Mediterranean in particular, and they therefore stimulated and shaped the production of translations. Under such circumstances, translation could be a call for visibility and a way to make a living. Yet, translation could also be a transgressive experience that further shadowed the visibility of enslaved people. An enslaved women from Belgrade, for instance, assisted late seventeenth-century Breslau pastor Andreas Acoluthus in the production of his famous Arabic, Persian, Ottoman and Latin Qur'an specimen. She had been the wife of a Belgrade imam prior to her captivity, and she ran away only a few months after her arrival in the household of the Lutheran scholar even though he had emancipated her. As I argued elsewhere, such a story 'hints at her possible experience of such a linguistic encounter as a troubling transgression of personal, religious, and gender boundaries'.[11] Like so many women translators of the time, her full biography still waits to be written; and this in spite of the fact that women, as Peter

8 Kate Briggs, *This Little Art* (London: Fitzcarraldo Editions, 2017); Theo Hermans, *The Conference of the Tongues* (London: Routledge, 2007).

9 Lawrence Venuti, *The Translator's Invisibility: A History of Translation*, 2nd ed. (London: Routledge, 2008).

10 John Gallagher, 'Migrant Voices in Multilingual London, 1560–1600', *Transactions of the Royal Historical Society* 2 (2024): 39–61.

11 Stefan Hanß, 'Ottoman Language Learning in Early Modern Germany', *Central European History* 54, no. 1 (2021): 1–33 (16). See also Alastair Hamilton, 'A Lutheran Translator for the Quran: A Late Seventeenth-Century Quest', in *The Republic of Letters and the Levant*, ed. Alastair Hamilton, Maurits H. van den Boogert and Bart Westerweel (Leiden: Brill, 2005), 214; Alastair Hamilton, '"To Rescue the Honour of the Germans": Qur'an Translations by Eighteenth- and Early Nineteenth-Century German Protestants', *Journal of the Warburg and Courtauld Institutes* 77 (2014): 173–209, 177–179.

Burke has stressed, 'were relatively prominent in this field, probably because translation was considered more compatible than original writing with female modesty'.[12] While the importance of women for the kinshipping prospects of translator and interpreter families is well documented, a fuller history of early modern women translators has still not been written.[13]

The Mediterranean lens of many of the contributions to this volume, moreover, presents an important theoretical angle that resonates with wider reflections in the field of Translation Studies. Translators themselves, like Mireille Gansel, call attention to the 'transhumance routes of translation, the slow and patient crossing of countries, all borders eradicated, the movement of huge flocks of words through all the vernaculars of the umbrella language of poetry'.[14] Now, this matters since transhumance was a practice of mobility widely shared across the early modern Mediterranean, and one that (travelling or residing) translators were widely confronted with.[15] Moreover, Translation Studies increasingly approach early modern translations as mediated through travels. Guyda Armstrong for instance puts emphasis on translation 'as products of a geographically situated practice', and has called for 'a much more decentred, multiple, multilingual, horizontal and, most of all, *mobile* model of how literary texts actually function and are made in the period'.[16] This volume's focus on the 'context of production' of translation, as Alessia Castagnino and Giovanni Lista call it in their contributions, is thus a powerful means to provide more multi-scalar stories of historically contextualised, situated translations.

---

12 Peter Burke, 'Cultures of Translation in Early Modern Europe', in *Cultural Translation in Early Modern Europe*, ed. Peter Burke and R. Po-chia Hsia (Cambridge: Cambridge University Press, 2007), 7–38, 12. For example, see Madeleine de l'Aubespine, *Selected Poems and Translations: A Bilingual Edition*, edited and translated by Anna Kłosowska (Chicago: University of Chicago Press, 2007).

13 On the role of women in dragoman families, see E. Natalie Rothman, *The Dragoman Renaissance: Diplomatic Interpreters and the Routes of Orientalism* (Ithaca: Cornell University Press, 2021). For early modern England and Germany, see Deborah Uman, *Women as Translators in Early Modern England* (Newark: University of Delaware Press, 2012); Hilary Brown, *Women and Early Modern Cultures of Translation: Beyond the Female Tradition* (Oxford: Oxford University Press, 2022). For the modern period, see Luise von Flotow, ed., *Translating Women* (Ottawa: University of Ottawa Press, 2011); Luise von Flotow and Farzaneh Farahzad, eds., *Translating Women: Different Voices and New Horizons* (New York: Routledge, 2017).

14 Mireille Gansel, *Translation as Transhumance*, translated by Ros Schwartz, foreword by Lauren Elkin (London: Les Fugitives, 2017), 77.

15 Braudel, *The Mediterranean*, 1, 85–102; J.R. McNeill, *The Mountains of the Mediterranean World: An Environmental History* (Cambridge: Cambridge University Press, 1992).

16 Guyda Armstrong, 'Place and Space in Mediated Translations', *Forum for Modern Language Studies* 58, no. 4 (2022): 497–504 (498–499).

While the historian's focus on cultural contexts will be of pivotal significance for historicising early modern translations, this volume also recovers the ways in which experiences of spatial mobility impacted translators' works. Translators 'are never ... neutral, impersonal transferring devices', writes Douglas Robinson. 'Translators' personal experiences – emotions, motivations, attitudes, associations – are not only allowable in the formation of a working [translation], they are indispensable', and it is thus the historian's task to recover such personal experiences in the past.[17] Unsurprisingly, translators themselves have insisted that the experience of travels is pivotal in the making of translations.[18] Now, if translation is 'a *site* for learning' and thus itself a spatial terrain of doing and thinking, a focus on translation also entails important learning experiences for historians, in particular the invitation to reflect on the ways we write the history of the early modern world.[19] This volume calls for more mobile, unstable, and reflective stories of cultural exchange and transfer, and stories that constantly require historians to reconsider and shift the different perspectives involved in language encounters. Thinking about languages and translatability, in Natalie Zemon Davis's words, is essentially an exercise in 'listening'.[20]

So, what exactly can we listen to? I would insist it is the 'translation possibilities'[21] that arise from past translators' 'creative decisions and imaginative acts of criticism'.[22] The momentum of agency and creativity, in other words, invites historians to recalibrate contexts, to de- and recontextualise translators' lives as a continued exercise that itself resembles the ways translation was done.[23] Translation provokes 'thinking [about] relationality on a micro-scale', which is at the heart of the work of early modern historians.[24] Translation can also cause

17 Douglas Robinson, *The Translator's Turn* (Baltimore: The Johns Hopkins University Press, 1991), 260.

18 Gansel, *Translation as Transhumance.*

19 Briggs, *This Little Art*, 210.

20 Natalie Zemon Davis, *Listening to the Languages of the People: Lazare Sainéan on Romanian, Yiddish, and French* (Budapest: CEU Press, 2022).

21 Briggs, *This Little Art*, 192.

22 Edith Grossman, *Why Translation Matters* (New Haven: Yale University Press, 2010), 8.

23 On the ways that a focus on translations changes the ways we do history and cultural studies, see Doris Bachmann-Medick, 'Introduction: The Translational Turn', *Translation Studies* 2, no. 1 (2009): 2–16; Lawrence Venuti, *Translation Changes Everything: Theory and Practice* (London: Routledge, 2013); Doris Bachmann-Medick, *Cultural Turns: New Orientations in the Study of Culture*, translated by Adam Blauhut (Berlin: De Gruyter, 2016).

24 Kate Briggs, *This Little Art* (London: Fitzcarraldo Editions, 2017), 160 (quote). See also Peregrine Horden and Nicholas Purcell, *The Corrupting Sea: A Study of Mediterranean History* (Oxford: Blackwell, 2000); Stefan Hanß and Dorothea McEwan, eds., *The Habsburg Mediterranean, 1500–1800* (Vienna: Austrian Academy of Sciences, 2021).

a sense of confusion, disruption and unfamiliarity that is all too familiar to historians stumbling over a puzzling archival find that prompts us to reflect on positionality and perspectives, and helps develop critical thinking to handle, approach and make sense of such finds. Not the least, translation reminds historians of what words can do, or as Peter Burke once aptly put it: 'even the most monoglot of historians is a translator'.[25] The historian of the early modern world, one might conclude, cannot be anything other than a translator who is attuned to the lives of translators in the past.

---

25 Burke, 'Cultures of Translation', 7.

Index

www.ingramcontent.com/pod-product-compliance
Lightning Source LLC
Chambersburg PA
CBHW071333230926
51849CB00002B/382
* 9 7 8 9 0 0 4 7 3 1 7 9 0 *